"Packed full of scientific insights with practical applications to everyday life—a thought-provoking and entertaining page-turner."

—Gary Small, MD, UCLA professor of psychiatry and author of
The Memory Bible, iBrain, and *The Other Side of the Couch:
A Psychiatrist Solves His Most Unusual Cases*

"Reading *What Makes Your Brain Happy* is like eating intellectual dim sum at your favorite Chinese restaurant. Each morsel is tasty, and you will keep coming back for more."

—Bruce Hood, PhD, author of *SuperSense:
Why We Believe the Unbelievable* and director of
the Bristol Cognitive Development Centre

"DiSalvo takes us on a refreshing voyage into the multitude of ways your brain is busy smacking you around, and provides an antidote to the standard servings of self-help snake oil."

—Mark Changizi, PhD, author of *Harnessed: How Language
and Music Mimicked Nature and Transformed Ape to Man*

"The chapters in this book are crystal clear and multifaceted, and each transmits a ray of insight about how we think. It's jewelry for the mindful mind."

—Phillip Alcabes, PhD, author of *Dread: How Fear and Fantasy
Have Fueled Epidemics from the Black Plague to the Avian Flu*

"It's hard to put down this smart, readable discussion of the latest brain science. . . . As always, DiSalvo deftly offers both expert and lay readers news we can use, in context and with style. Read on!"

—Maggie Jackson, author of *Distracted: The Erosion of Attention
and the Coming Dark Age*

"DiSalvo is a genial and enthusiastic guide who makes emerging research in neuroscience, social psychology, cognitive science, and behavioral economics accessible, and even entertaining. But this book is not specifically about research, nor is it really about brains and minds. What it is about is you and me and how science can help with the messy business of trying to live a meaningful, good life. A delightfully illuminating read."
—Todd Essig, PhD, training and supervising psychoanalyst,
William Alanson White Institute

"Every week the media delivers to the public a barrage of psychology and neuroscience findings. They sound fascinating but are untethered from daily life. DiSalvo extracts the practical potential of these discoveries, and in so doing performs a public service that is creative and witty."
—J. D. Trout, PhD, author of *The Empathy Gap: Building Bridges to the Good Life and the Good Society* and professor of philosophy and psychology, Loyola University Chicago

"Want to know how all the little ins and outs of your neurochemistry hold you dangling like a puppet on a set of strings? DiSalvo's gleanings from current neuroscience and psychology are entertaining, intriguing, and instructive. And who knows? The more you know, the more control you might gain over how those strings make you dance."
—Joseph Carroll, PhD, author of *The Literary Animal: Evolution and the Nature of Narrative*

"DiSalvo's book provides a well-written foray into the fascinating fields of neuroscience and social psychology. It will pique your curiosity and help you understand people in new ways."
—Charles H. Elliott, PhD, coauthor of many psychology books, including *Overcoming Anxiety for Dummies*, *Obsessive Compulsive Disorder for Dummies*, and *Borderline Personality Disorder for Dummies*

A Revolutionary
New Weight Loss Plan
to Lower Your
Breast Cancer Risk

The Pink Ribbon Diet

Mary Flynn, PhD, RD, LDN
and Nancy Verde Barr

Da Capo
∞
**LIFE
LONG**

A Member of the Perseus Books Group

Copyright © 2010 by Mary Flynn and Nancy Verde Barr

Editorial production by *Marra*thon Production Services. www.marrathon.net

Design by Jane Raese
Set in 9.5 point Centennial

Library of Congress Control Number: 2010928400

ISBN 978-0-7382-1394-1

First Da Capo Press edition 2010

Published by Da Capo Press
An imprint of Perseus Books, LLC
A subsidiary of Hachette Book Group, Inc.

Da Capo Press books are available at special discounts for bulk purchases
in the U.S. by corporations, institutions, and other organizations.
For more information, please contact the Special Markets Department at the
Perseus Books Group, 2300 Chestnut Street, Suite 200, Philadelphia, PA 19103,
or call (800) 810-4145, ext. 5000, or e-mail special.markets@perseusbooks.com.

20 19 18 17 16 15 14 13 12 11 10

Breast cancer heroes come equally

from those who battle with the disease

and those who serve to eliminate it.

This book is dedicated to all those brave individuals

who are waging this gravest war.

And especially to Lhasa . . .

"some clouds leave great tracks in the sky . . ."

—*Cloud*, The Barr Brothers

Contents

Preface

I told Mary that I wanted to write my introductory text first, even though the material in this book is hers. The Pink Ribbon Diet is her diet, developed from her own study of breast cancer patients and weight loss. So as a cowriter, why did I want to jump in before her? Well, as they say, none are so vocal and fervent as the converted.

Food has been my passion and my profession for more than thirty years. I spent eighteen of those years working as a culinary associate to Julia Child, a food lover but a skeptic of food as medicine. Mea culpa. I had shared her view: The gourmet cares about taste; the dietitian does not. And then I met Mary Flynn.

I met Mary through our work with the Rhode Island Community Food Bank. When the local PBS TV station asked me to host a series of short programs aimed at improving nutrition for families with children who were at risk due to obesity, I realized that I needed a nutrition expert, so I asked Mary to join me on camera to give the facts. The day she brought her granola roasted in olive oil with almonds and raisins to the set and the cameramen couldn't stop eating it, I realized that this dietitian cared a lot about taste. She brought more samples of her cooking to a Food Bank meeting, and my previously held opinion that dietitians can't cook was history.

Mary cares about good food. Most importantly, her work with patients suffering from chronic disease made her examination of the nutritional benefits of food critical.

After an intensely stressful few weeks last year, I developed health problems of my own: inexplicable daily stomachaches. My physician prescribed a clear-liquids-only diet. After a few days, I was miserable and very hungry, but when I tried to introduce solids, the pain returned. I happened to mention this to Mary, and she said I could try certain foods, such as white rice and eggs cooked in olive oil, plain mashed baked potato, even a glass of red wine. After following her advice for a few days, I could soon tolerate other foods and all the pain was gone. When I asked Mary what it was about those foods that worked so well, she gave me a short course in how some foods can help alleviate disease. Hmm, I thought. Tell me more. Mary then described her work with overweight breast cancer patients and how the foods she prescribed helped improve their chances of survival. The positive results of her study, which was sponsored by the Susan G. Komen for the Cure Foundation, were undeniable.

Courses in nutrition are a requisite for culinary training, so I knew a fair amount about the subject before I met Mary. I ate plenty of vegetables and whole grains and kept my intake of red meat to a minimum. That's enough, right? As it turns out, it's not even close. I had only part of the story. I didn't know enough about how certain foods react with each other, which nutrients are needed for which body functions, or why some cooking methods rob foods of their nutritional value. And I didn't know that certain foods have been linked to either increased or decreased risk of breast cancer. Mary filled in the missing pieces for me.

Shortly after Mary told me about the Komen study and her work with breast cancer survivors and diet, a special friend of my sons—a beautiful, young, talented musician—was diagnosed with advanced stage breast cancer. Her desperation to find answers and some sense of control pained my boys, and they asked me what I knew about diets for cancer patients. Immediately, I thought of Mary's work—and realized how much not only a woman struggling with breast cancer but also her family and friends need immediate access to sound, unassailable information. I suggested to Mary that she share her study and her diet in a book.

The Pink Ribbon Diet is a book for all women—whether you are facing a breast cancer diagnosis, are recovering from cancer, or want to minimize your risk of developing the disease. Mary built a flexibility into her diet that makes going—and staying—on it easy. Although her personal diet does not include poultry or red meat (see recommendations on page 36), she didn't blink when I suggested chicken and turkey recipes for the book. Her goal is to make me, and all her patients, realize the healthiest diet possible within the confines of what we can live with. I doubt that anyone, myself included, looks forward to dieting, but her plant- and olive oil–based Mediterranean-style eating plan does not feel like a diet. Her eating plan has plenty to eat and, most importantly, plenty of flavor.

Throughout this book, you'll find stories from women, many of them breast cancer patients, who have successfully lost weight on this diet and found a renewed sense of well-being. Before beginning to work on this book with Mary, I started her diet. After two weeks of following her plan religiously, I was amazed at my own increase in energy. My complexion and hair looked great. When my sons saw me after those two weeks of the plant-based olive oil diet (PBOO) eating, they remarked that I looked ten years younger. Wow!

So how long did I follow the Pink Ribbon diet? I still do—because it is about not so much restricting foods in your diet but adding great foods packed with major health benefits: olive oil, vegetables, whole grains. The diet really is a piece of cake—and yes, you can have cake, cookies, or pie for dessert. What's not to love? Eating delicious, cancer-fighting food, losing pounds, having more energy, and feeling ten years younger. I repeat—Wow!

I recommend this diet to anyone who wants to improve their health and especially those who need to lose weight—why not eat to accomplish both?

Nancy Verde Barr

Introduction

One in eight women will be diagnosed with breast cancer in her lifetime. In 1960, the numbers were one in twenty. Despite advancements in early detection and treatment, when it comes to breast cancer prevention, it seems—to paraphrase a song lyric—"For all the miles we've put behind/ we haven't come that far.*

During the last fifty years, we've actually gone backwards, even though we know that one of the primary contributors to increased risk of breast cancer is excess weight. It's not just me saying this! Published research from around the world confirms the relationship between being overweight and the likelihood of developing the disease.

Although scientists have promising theories about the link between weight and breast cancer, they do not yet fully understand the mechanics. However, they are certain that the more body fat you have, the higher your cancer risk. Overweight, postmenopausal women and young women who gain weight from eighteen years old throughout adulthood are in particular danger. They have a 10 to 35 percent greater risk compared to women of normal weight. That risk skyrockets to a frightening 56 to 82 percent for women who are not only overweight but obese.[†1]

Consider this information in light of an equally alarming statistic from the Centers for Disease Control and Prevention (CDC): 67 percent of adults, well more than 100 million Americans, are overweight or obese. Breast cancer specialists have good cause to be concerned about America's "fat epidemic."

Everyone can lose weight. I want to emphasize that: Everyone can lose weight. For many of you, that statement may well elicit uncomfortable

---

*The Barr Brothers, "The Devil's Harp."

†Your weight status is determined by the relationship of weight to height, or body mass index (BMI). If you have access to the Internet, go to http://www.cdc.gov/healthyweight/assessing/bmi/index.html to determine your ideal BMI. Otherwise, ask your physician to calculate it for you.

1

feelings that range anywhere from a gloomy sense of defeat to an angry, resentful challenge: "Well I can't and I know it!" I get it.

For twenty-five years, I have counseled clients and patients who felt the same way and had the same reasons you probably do: "I've tried so many times and I always fail, so what's the use?" "It's too hard." "I like to eat too much." "I have to feed my family and they don't like diet food." "Donuts just find me." And so on and so forth. I've heard an impressive host of reasons and still, with total conviction, I'm telling you, "Everyone can lose weight—especially you!"

Because you are reading this book, most likely you or someone you love is going through or has been through a life-threatening struggle with breast cancer. Perhaps you, like the women you will read about in this book, have gained weight during treatment. The statistics for breast cancer patients who put on weight are ugly. A gain of as little as 2 to 4.4 pounds can increase the risk of cancer recurrence and death by 35 percent; more than 4.4 pounds raises that risk to a grim 65 percent.[2]

Perhaps you picked up this book not because you have had breast cancer but because you have heard strong and sobering evidence connecting excess weight to breast cancer. If you are one of the 67 percent of overweight Americans, you are at risk—even more so if you are postmenopausal or a young woman who has gained weight from eighteen years old to adulthood. If breast cancer runs in your family, your risk is greatly increased and your motivation to lose those excess pounds and maintain a healthy weight should be overwhelming.

People with compelling reasons to be motivated—looking good for those class reunions, fitting into a dress for an upcoming wedding, an invitation to a pool party—are more successful at losing weight than those who do not have a motivating goal. What could be more motivating than defending yourself against a battle with illness that takes so much from you and your family? Chances are, you have had some success losing weight in the past but gave up for any number of reasons. At this time in your life, giving up is not an option.

For those who have survived breast cancer, staying cancer-free is a powerful motivator but not necessarily the only one. After invasive breast cancer, many women find that their self-esteem has been injured. Weight gain is insult added to injury. You want to be able to look in the mirror and not be reminded that you have had breast cancer; you want to be able to say that you never looked or felt better.

Successful weight loss does require change, including a change in how you think. In the words of Winston Churchill, "Attitude is a little thing that makes a big difference." To many of you, diet means deprivation, misery, even torment. I am asking you to change that way of thinking. The eating plan in this book is not a punishment. It is an opportunity to be more—or in terms of body size, less—than you ever thought you could be.

How can I be so sure that you can do this? The women I worked with in my study, which you'll read about in the following pages, were just like you. They had struggled for years with their weight and finally found the answer in my diet. I want you to feel that same sense of control as those women, whether you have breast cancer, are at risk for breast cancer, or just want to lose weight while optimizing your health. With that goal in mind, I brought their program to you in book form, *The Pink Ribbon Diet.* Don't spend another day lamenting extra pounds, agonizing about your well-being, or worrying about the illnesses those pounds might cause. Take the first step now to lifelong control over your weight.

In 2004, I was awarded a grant by the Susan G. Komen for the Cure Foundation to initiate a study at The Miriam Hospital in Providence, Rhode Island. My objective was to examine the effects of diet on women who had completed treatment for invasive breast cancer and were overweight, a situation that invites a high risk of recurrence. I had two goals: One, to compare the low-fat diet recommended by the National Cancer Institute (NCI) to a Mediterranean-style diet that I had developed—a plant-based olive oil diet (or PBOO diet)—to determine which one was more effective in improving some of the **biomarkers** for breast cancer (that is, lab test results that are related to the risk for the disease). And two, to determine if my higher-fat diet, with its emphasis on extra virgin olive oil, would be as effective for weight loss as the reduced-fat diet recommended by the National Cancer Institute.

I had good cause to pit my diet against low-fat diets such as the one prescribed by the NCI.* When I was examining the research for my PhD dissertation on low-fat diets and blood fats, I found study after study on

---

*In the 1980s, the NCI was the first agency to promote low-fat diets. Believing that there would be a large decrease in the incidence of breast cancer if women adopted a low-fat diet, the NCI put a great deal of money into research. However, after researchers failed to find a connection, the NCI modified its recommendation to read "go easy on fat." The diet as outlined by the NCI remained and still is a low-fat one.

the frequent failure of low-fat diets and their health risks. Armed with extensive and captivating information, I continued what became twenty-five years of researching diet and chronic disease (and teaching at Brown University). My research repeatedly led to the same conclusion: All high-fat diets do not lead to weight gain and are not linked to increased risk of breast cancer. Quite the contrary; the right fats in the right amounts are crucial to optimizing health and beneficial to shedding pounds. The "right fats" are healthy fats, and only extra virgin olive oil can improve health.

After I wrote the grant for my study, other studies were published that supported what I was doing; that is, they failed to prove that a diet low in fat is effective in fighting breast cancer. The biggest flop was the Women's Health Initiative, published in 2006, which studied forty-nine thousand women, cost $415 million, and determined that a low-fat diet did not prevent breast cancer.[3] Oops!

On the other hand, an analysis of five thousand cases of breast cancer found that women whose diets contained less than 15 percent dietary fat doubled their risk of breast cancer.[4] Big oops! But it gets worse. Low-fat diets can rob the body of nutrients needed to defend against cancer, rendering some otherwise sound cancer-fighting advice useless. For example, the NCI wisely advocates a diet rich in carotenoids. **Carotenoids**, a family of **phytonutrients** found in plant products, give carrots, squash, broccoli, corn, tomatoes, and other vegetables their bright color and have been shown in many studies to reduce the risk of breast cancer.[5] To absorb these vital nutrients, however, our bodies need fat.[6] Eat your vegetables without it and all those lovely little carotenoids with their impressive cancer-fighting capabilities are simply wasted.

Equally embarrassing for proponents of a low-fat diet is that researchers have been unable to come up with any proof that it is effective for weight loss. In 1998, Walter Willet, chairman of the Department of Nutrition at the Harvard School of Public Health, surveyed the scientific literature on dietary fat and concluded, "Diets high in fat do not appear to be the primary cause of the high prevalence of excess body fat in our society, and reductions in fat will not be a solution."[7] Try telling that to food manufacturers who have put, if not their heart and souls, at least their money and public relations staff into convincing us that the no fat, reduced-fat, and low-fat packaged foods littering the grocery shelves are the answers to a dieter's prayer. And that is the problem: We have been binging on low-fat lies. "For almost thirty years, Americans have been cutting back on fat, or trying to. The percentage of fat in the American diet has dropped by almost 15 percent. And all this time, Americans have continued to get fatter—over 30 percent fatter."[8] Really big oops!

But you don't need to be discouraged. The answer is not to eliminate fat from your diet but to replace unhealthy fats with healthy olive oil. The high healthy-fat diet, which I developed for overweight women with breast cancer, did what it was designed to do. The women lost weight and their biomarkers improved. An Israeli study released after my study was completed supports my findings. Published in the *New England Journal of Medicine,*[9] the study compared a low-fat diet with a low-carb diet and a Mediterranean diet. At the end of the two-year study, those on the low-fat diet lost an average of 5.5 pounds while the Mediterranean group lost 10 pounds and low-carb dieters lost 10.3 pounds. But, and this is a very big but for all females, the women in the study tended to lose much more on the Mediterranean diet—an average of more than 13 pounds compared to about 5 pounds on the low-carb and less than 1 pound on the low-fat diet.

For readers of this book, the significance of this study is that it determined that the higher fat, olive oil–based Mediterranean diet had measurable health benefits not exhibited with the low-fat diet: namely, improvement in measurements of blood lipids, blood sugar, and **insulin**—all biomarkers for breast cancer.

At the risk of sounding like that annoying student in the front row with a hand always raised—I knew that! Like the successful diet of the Israeli study, my PBOO diet is based on the healthful, Mediterranean eating plan. But the PBOO diet is breast cancer specific in that it omits foods linked to increased risk and is weighted with foods containing nutrients that improve biomarkers and are considered to lower risk. Perhaps you're thinking that if we know what those miracle nutrients are, why not just isolate them in pill form and wash them down daily with three cans of Slim Fast? Not so fast. Turns out, researchers and nutritionists have already tried that—well, not the drink, but the pills. It doesn't work.

According to Christopher Gardner, nutrition scientist at Stanford Prevention Research Center, "The frustrating thing in nutrition is that for the last couple of decades, so many studies have failed because we've isolated one nutrient at a time, when the benefit comes from the synergistic and additive effects of the diet taken together."[10] That certainly warrants restating. From Linda Van Horn, acting chair of preventative medicine at Northwestern University's Feinberg School of Medicine: "Rather than reducing a diet to its essential nutrients—vitamins, minerals, and other chemicals—and trying to isolate those that contribute to health, researchers are increasingly taking a step back and correlating health with broader eating patterns."[11]

The diet I developed for the women in my program does just that—it incorporates the foods that contain those "miracle nutrients" into the healthy pattern of the Mediterranean diet. Curiously, although much has been published in medical journals about the increased risk of breast cancer caused by excess weight and in spite of the stupefying number of women whose weight puts them at risk, precious little in the way of cookbooks addresses the complete situation. Losing weight is only half the battle; the other is improving those crucial measurements of risk.

Anyone who has heard the alarming diagnosis "It's breast cancer" knows the desperation of searching for answers. Magazines, the Internet, bookstores, and the media are rife with information about causes and cures. Unfortunately, a great deal of that information is incomplete or unsubstantiated. Headlines may make compelling arguments; the science behind them is often less so. Out of necessity for my teaching position at Brown, as well as my own interest, I try to read all the studies associated with the diseases I teach. When one study reports a link between a certain food and breast cancer, I am interested. But I wait. When several more reports support that study or find no connection, I pay attention. I consider the size of the study, its source, its conditions, and whether the study was conducted with humans or animals. I dig back into twenty-five years of research and clinical experience. Only after a thorough examination do I come to my own conclusion. Consequently, everything I included or omitted from my diet was based on a good deal of substantiated scientific study.

By any yardstick, my diet study was a success. The women in the program lost weight, their biomarkers improved, and their energy increased. Equally satisfying to me was that they loved the diet and heartily recommended it to others. And, as it turned out, so did physicians at three local hospitals who, after seeing the results of the diet and hearing their patients' enthusiasm for it, referred their private patients to me.

I started the program at The Miriam Hospital as a scientific study to measure and compare weight loss and biomarkers. But when I look at what my program meant emotionally to the women I helped, the most rewarding results are immeasurable. I could not have predicted, nor can I measure, the boost to their self-confidence. Women who had been through tough battles with breast cancer and excess weight finally felt in control of their lives.

Mary Flynn, PhD, RD, LDN

Part One
The Diet

Battling Two Foes

One must not forget that recovery is brought about not by the physician but by the sick man himself. He heals himself, by his own power, exactly as he walks by means of his own power, or eats, or thinks, breathes, or sleeps.

—Georg Groddeck, physician and writer,
The Book of the It (1923)

Fran was an energetic, super healthy, fifty-seven-year-old working wife and mother of two when she became a statistic. A routine physical exam revealed a lump that proved to be stage II-B breast cancer, linking her life to those of a staggering two and half million women living with this disease. In the words of Dr. Seuss, "I have heard there are troubles of more than one kind. Some come from ahead and some come from behind." Dealing with her trouble ahead—the shock, the fear, a difficult course of chemo, surgery, more chemo, and radiation—Fran did not anticipate the trouble that came from behind. She was rapidly gaining weight.

"I wasn't expecting it, but I wasn't distressed by it either. I zoomed all my mental energy into saving my life. I got used to being puffy and round and just kept buying stylish, bigger sizes. When I would get weighed in each week before chemo, I would lament on my gaining, but the nursing staff told me not to worry about my weight. I didn't, but when my treatment was over and my postsurgical pathology was the best I could hope for, I wanted to look as good as I felt. I wanted to get back to my precancer happy, energetic lifestyle, but my body looked worse than ever. Scars on my breast and under my arm were lovely compared to my waist, which was completely gone. My energy was at an all-time low from carrying around the extra pounds."

Time out. Weight *gain?* How can that be? Aren't gaunt and emaciated the common images of a cancer patient? Turns out, not always. According to the American Cancer Society, over half the women undergoing treatment for breast cancer may gain rather than lose weight.[1]

The Health, Eating, Activity, and Lifestyle (HEAL) Study, which began in 1996 to examine—among other factors—the relationship between weight and breast cancer, looked at the cases of twelve hundred women across the United States. The findings were remarkable, to say the least: From the time of diagnosis through treatment, 74 percent of women battling breast cancer gained body fat and 68 percent gained pounds. The average weight gain was 8.6 pounds, with a range from less than 1 pound to 59 pounds.[2]

Talk about a fight—these women were battling breast cancer *and* the bulge.

> I lost my waist during treatment and
> haven't been able to find it since.
> —*Caroline*

If it's not one thing, it's the hormones. At least that's what most women blame for the uninvited and unwelcome pounds. But the data is inconclusive. The American Cancer Society (ACS) considered a number of possible causes of weight gain: Intense food cravings following the nausea of chemotherapy; slowed metabolism brought on by chemotherapy-induced menopause; decrease in physical activity due to the fatigue that accompanies therapy; and the aforementioned usual suspect, hormone therapy, which is often part of the treatment.[3]

ACS researchers found no definitive scientific facts pointing to any sole perpetrator. Not knowing what is causing the pounds to pile on is usually frustrating to women who are trying desperately to gain some control over their illness. As Lois remembered, "I was not told by my MD that I might gain weight once on the tamoxifen (now Arimidex) and was not told that the low-dose Effexor prescribed for the extreme night sweats would also cause weight gain. I lasted two weeks on Effexor, during which I could not button my pants or shirts—like someone took a bicycle pump to my body! Would rather not sleep. When I complained about gaining weight, the answer I got was that it was not the pills but the result of menopause. Since I was already in menopause for several years, I'm not sure I accept that."

Although the exact causes of weight gain remain uncertain, the effects do not. Excess weight increases blood insulin level, which is related to breast cancer risk and mortality. In addition, fat cells in the body make **estrogen,** and estrogen nourishes breast cancer. Although the medical repercussions of weight gain are sobering, the psychological and emotional threat to recovery it can cause should not be underestimated. For good or ill, most women link their self-confidence to how they look, and becoming fat after the loss of a breast or other physical changes is yet another blow to an already wounded self-image. Renee would tell you that there's little room for a healthy dose of self-confidence while trying to squeeze your body into a two-month-old dress that is already three sizes too small.

Renee was fifty-three when she was diagnosed with right breast infiltrating duct and lobular carcinoma. During her treatment, she put on almost forty pounds. She still finds it difficult to talk about her weight gain. "When I went to the doc's office, my weight was measured and I was gaining. It seemed all of a sudden to increase very fast. The doc seemed to take the weight gain lightly, told me not to worry [and] that it would level off, but it never did. I kept gaining and gaining. Before cancer, I was a size ten and weighed 142. At my highest point during treatment I weighed 181 pounds and needed a size-eighteen dress! My mom would ask if I was losing any weight and I would tell her 'no' and get angry, but I hated the way I looked. I felt so unattractive."

Self-confidence and a positive attitude are vital to the can-do approach that so many cancer patients attribute to their survival. What made this wormy can of weight gain particularly problematic to these women in their struggle to win back their lives was that few of them were told by their physicians that weight gain was to be expected. Unaware that weight gain is common, most women felt solely responsible for the extra pounds and isolated in their struggle to lose them. Most simply could not. Without adequate information and dietary counseling from their health care professionals, they failed to lose any weight that they had gained and many felt defeated and depressed.

Cheryl spoke of her escalating depression. "During treatment, I believe I was depressed but didn't fully understand it because I thought I was superwoman and could handle anything. Normally always on the go, I fought the boredom of being out of work, the sadness of hair loss, and the esophagitis from chemo by eating myself through the year of treatment, watching *Ellen* and *Oprah* almost every day. I would eat breakfast, then a snack, then lunch, then a snack, then dinner. Eventually, I gained more than twenty pounds and felt so miserable."

> During my whole treatment, my weight was trending upward.
> I started buying clothes with elastic waists and sensible shoes.
> Elastic waists and sensible shoes are very depressing.
>
> —*Mimi*

Sadly, even when the women asked for help, some received none. A routine mammogram revealed the lump that a biopsy would prove was a grade 1 (slow-growing) tumor for fifty-eight-year-old, postmenopausal Mary-Ellen, whose weight burgeoned to 172 pounds during treatment. Determined to bring her own "self-healing energy into the fight," she was cancer free after nine months of treatment. Then she began her fight to stay that way. "I began to read magazine articles, go on the Internet, and pick up booklets at the doctor's office. The articles emphasized that researchers were leaning towards excess weight and persistent weight gain after cancer as a strong factor in breast cancer recurrence."

Mary-Ellen made an appointment with her primary care physician, but she was away so a partner in the practice saw her. "This female doctor examined me, and then told me two things, and I quote: 'You are obese and you must lose weight, and you are depressed and you must get some help for that.' That was the total extent of her advice, help, or support. After that, I refused to ever see her again." Of course Mary-Ellen was depressed! She had battled breast cancer and became fat in the course of doing so.

The physician should have informed Mary-Ellen that weight gain often follows cancer treatment and then referred her to a dietitian. Instead, the doctor's comment that "you are obese" made Mary-Ellen feel as if the entire fault lay with her, adding to her depression. When Mary-Ellen saw her own physician, she also recommended that she lose weight, suggesting that she "cut down on sweets and increase her exercise." It was not enough. Without specific guidance and ongoing support, Mary-Ellen found it increasingly difficult to shed pounds. Obviously, something is seriously wrong with this picture.

The Komen Study

I met Fran, Cheryl, Lois, Renee, and Mary-Ellen when they enrolled in the Komen study that I started at The Miriam Hospital in Providence, Rhode Island to compare the low-fat diet recommended for women with breast cancer by the National Cancer Institute to my plant-based, higher fat—in the form of extra virgin olive oil—diet. I believed that my diet would be as effective—if not more so—for weight loss and improvement of biomarkers as the diet from the NCI. My diet emphasizes foods considered to lower breast cancer risk and foods that improve biomarkers, such as olive oil, vegetables, and certain starches, and omits those foods linked to increased risk. But more importantly, my diet is a pattern for eating, a blueprint designed for the overweight breast cancer patient. These women weren't just dieting due to an upcoming wedding or bathing suit season; they were saving their lives. Just *any* diet wouldn't do!

Because I couldn't test my diet on women who had no weight to lose, I advertised for women who were overweight and had completed their cancer treatment. As I screened the more than 150 women who applied, I was startled to learn that all but one had gained weight during treatment. My surprise at the high incidence of weight gain paled in comparison to my frustration at how many of these women reported that they received little or no nutritional advice from their physicians, even when they asked. These women were not alone. A study published in the *Journal of Oncology*,[4] which addressed weight gain after diagnosis, reported that 52 percent of breast cancer survivors wanted nutrition guidance but few received any from their physicians. Wouldn't you think these physicians would be the very people who would know what kind of dietary guidance the women needed? Actually, they aren't. Most physicians will readily admit that nutrition is scarcely taught in medical schools. They had no advice to give other than to eat less, cut down on fats, and avoid sweets—none of which is all that is needed to improve breast cancer biomarkers.

I changed oncology doctors. I told my new doctor that I wished I
was not gaining so much weight, and she said the Komen study
would help me. God looked down on me that day!
—Renee

What is truly sad and frustrating is that these were women desperate
to gain some control over their lives. In her inspiring book and the sub-
sequent movie *Why I Wore Lipstick to My Mastectomy,* Geralyn Lucas
beautifully expresses the need that breast cancer patients feel to take
charge of their lives. In Lucas's case, taking charge was epitomized by
her wearing red lipstick—she had the tube of lipstick and knew where to
get more. To the women in my program, taking charge meant getting
their weight under control, but no one was giving them a diet plan and
they didn't know where to get one.

Lois was fifty-four when she was diagnosed with stage III, HER-2 positive breast cancer.

Having breast cancer put me on an emotional roller coaster. I felt ...
[that] the only things I could control which might impact favorably on
my health ... [were] my attitude, activity level, and diet. My primary
concern was to find the healthiest diet I could adopt. My feeling was, if
I could maintain excellent health (good cholesterol, blood pressure,
etc.) through diet and exercise, I would be in a better position to deal
with any future health problems that might arise. I did ask my doctors
what I should or shouldn't be eating but was unable to get enough information to change my diet. I attended a seminar at the Dana Farber
Cancer institute in Boston called "Fighting Cancer with a Fork." It was
helpful and provided recipes and tips but not a daily meal plan that I
could follow. The Komen study was just what I was looking for: a
dietary approach specifically designed for women who had breast
cancer.

The twenty-eight women I enrolled in the program began with the
NCI diet or my diet, followed the diet for eight weeks, and then switched.
The NCI diet for women diagnosed with breast cancer, as prescribed at
the time I wrote my grant, emphasized fruits, vegetables (five or more a
day), breads, and cereals; allowed 6 to 7 ounces of lean meat or fish a

day; and recommended lowering salt, sugar, alcohol, fats, and smoked or pickled foods. Otherwise, the diet was unspecific and nonstructured. I allowed the women following the NCI diet to choose from the recommended foods, in keeping with the diet's guidelines, but to keep within 1500 calories per day, which was the same allowance for my diet. Fat calories were limited to 2 tablespoons per day—the average amount consistent with a low-fat diet. I did create meal plans and recipes for the NCI diet so that the women would have a basic guide to follow.

My own plant-based olive oil (PBOO) diet was structured and specific. The women were given a diet chart (see Chart 2, "The Diet Grid," page 47) and told to choose a given number of servings per day of each category. Fruits and vegetables were not lumped together but listed in separate categories because I wanted the women to eat less fruits than vegetables (which are considerably lower in calories per serving than fruits). I allowed four to five servings per day of healthy fats, three of which had to be extra virgin olive oil. I asked that they consume a portion of the fat allowance at every meal. During the study, participants were allowed a total of 12 ounces per week of poultry or seafood.

Mary-Ellen began with the low-fat diet. "The diet was okay, but it was just regular food, including meat, chicken, and fish, with an emphasis on portion control and no sugar or alcohol. I did lose some weight but wasn't enthusiastic about the method."

When it was time for Mary-Ellen to switch to my diet, she doubted that it would work for her.

While it [the PBOO diet] sounded delicious, it looked to me as though I would have to learn to cook all over again. I had no clue what some of the foods were or how to cook them. I don't eat fish, and beef was less and less of an option, so I really felt that I was going on an almost strictly vegetarian diet. I thought that buying all that fresh food would be expensive and worried that the cost of the olive oil would be prohibitive when the study no longer provided it. I worried that I would be *hungry* all the time! Well, I thought, you beat cancer, you can do this for eight weeks.

Imagine my surprise to discover I loved the food! I loved the combination of vegetables, grains, pastas, and fresh seasonings, and the unusual salads. And it turned out that the foods were available in my local markets. To save money, I bought grains, rice, and cereal in bulk. I looked for sales on extra virgin olive oil. I've never spent so little on groceries in my life!

Cheryl, who was a lifelong member of Weight Watchers and had tried several different diets, also approached my diet with some skepticism.

It was different from the diets I had experienced in the past. It took some time to adjust to eating a lot more vegetables and adding extra virgin olive oil as well as more protein other than meat. For example, I had never tried beans or legumes other than a few, such as chickpeas on my salad. Much to my surprise, I started to really enjoy the wonderful flavors of olive oil with herbs and vegetables and beans. I preferred the plant-based diet to the low-fat one. Even my family started to enjoy several of the recipes that had been provided. A year has gone by and I have lost all the weight I gained during my treatment.

I was not surprised that the women in the program enjoyed the food in my high-fat diet. Fat adds flavor, and when that fat is flavorful *and* healthful, such as olive oil, you have a win-win situation. Furthermore, because fat consumed at a meal decreases and delays hunger between meals, olive oil contributed to the women's ability to maintain the diet.

The results of the study were impressive. As I predicted, the PBOO higher-fat diet was more effective than the NCI's low-fat diet both for improving biomarkers *and* for losing weight. From the time the women entered the program, they reduced their weight by an average of just slightly less than 16 pounds, lost 2.6 inches in their waists, and dropped their body fat percentages from 41.9 to 36.2. Most reported that their energy returned. Equally important, the plant-based diet was the one that the women found more satisfying and felt they could maintain for life. Women who had long struggled with their weight finally met with success—women like Mary Ann, who had tried hard to lose the weight she gained but failed.

I hated the weight gain and could not wait to deal with it. But when I tried to lose weight after treatment, I was unable to do so in spite of cutting down and going to the gym five times a week. I entered the

Komen study because I felt it was a great opportunity to work with a dietician and figure out how to eat. The plant-based diet was more successful than any I'd tried in that it was a slow, steady weight loss. Since the program ended, my weight has stayed the same or slightly lower.

Dr. Seuss had a solution for the troubles from ahead and those from behind—he "bought a big bat." The PBOO diet proved to be a very big bat for women in their battle against the two foes. Fran feels like her old self—only better. "I'm in control of the fight against a recurrence of the disease by PBOO eating and keeping active. It may not seem an important issue, but to me it sure is inspiring to continually hear 'You look wonderful.' My daughter-in-law says I have a different glow. I have a new energy level." Mary-Ellen is no longer depressed. On my diet, she lost 35 pounds and her cholesterol and blood pressure dropped. Three years later, she continues to follow the diet and has maintained her weight loss. "My doctors remain amazed at my blood pressure, cholesterol, triglycerides, and blood sugar. I recommend this diet to everyone." So do I.

Now I don't look at it as a diet. I consider it the right way to eat.

—Cheryl

What to Eat,
What Not to Eat,
and Why

If you are what you eat and you don't know what you're eating, do you know who you are?

—Claude Fischler, sociologist with the
French National Center for Scientific Research (2004)

I am as concerned about what you put in your body as what you keep out. You should be too. What you keep out may well make you lose weight, but what you put in can make your body the most efficient cancer-fighting machine it can be.

For generations, people living in the countries that border the Mediterranean Sea have thrived on a diet dominated by vegetables and whole grains prepared with olive oil and washed down with a glass or two of wine. For these populations, rates of chronic diseases are among the lowest in the world and life expectancy among the highest. So monumentally impressive was that finding that in 1993 the Harvard School of Public Health, the World Health Organization, and Oldways Preservation Trust, an internationally renowned nonprofit food think tank, gathered in Cambridge, Massachusetts, where they created the optimum food pyramid based on those obviously wholesome eating patterns. To this day, the Mediterranean Diet remains the gold standard of healthy eating.

The diet I developed for the women in my study is based on the same Mediterranean Diet food pyramid but is breast cancer specific. It eliminates any foods associated with the disease and emphasizes foods shown to have the ability to improve breast cancer biomarkers such as oxidation, inflammation, glucose, and insulin, which may be linked to the progress of your disease as well as the effects of treatment. You can improve the measurements of these biomarkers with a complete diet plan that includes the "right foods," described in detail in this chapter.

The "wrong foods"—the eliminated foods—are a no-brainer. If the food has been linked in a study to an increased cancer risk or is even suspect, it is banished. Out of here!

The following foods should be eliminated:

Red meat
Cured meats (such as ham, salami, or bologna)
Margarine (all kinds)
Vegetable oils (soybean, safflower, sunflower, corn)
Mayonnaise, unless made without partially hydrogenated fats and
    the main fat is olive oil or canola oil
Any food whose ingredients label lists "partially hydrogenated"
    (trans fats or **trans fatty acids**)

Check the foods in your pantry or on the grocery shelves carefully. The product may boast that it's healthy in outsized, front-of-the-package letters, but what does the nutrition label say? Is that mayonnaise or salad dressing made with vegetable oil? Skip it. Does that peanut butter or low-fat, low-cholesterol diet cookie or muffin contain partially hydrogenated oil (trans fats)? Toss them out.

Some foods are neutral; that is, they do no harm but do not necessarily improve biomarkers. In most cases, they are allowed or allowed occasionally. Condiments such as mustard, relish, vinegar, and soy sauce, for example, are neutral foods that add flavor but no or very few calories. They are allowed. Foods that are high in calories but contain none of the banned ingredients are allowed occasionally provided you have consumed the recommended daily allowance of required foods and the calories of the neutral food do not exceed your daily allowance. See Diet Grid on page 47 and read about calories on page 50.

The Big Three

The "right foods" are foods that can improve your health. The champions of this group are the "big three"—extra virgin olive oil, whole-grain starch, and vegetables. Chart 2 ("The Diet Grid"), on page 47, tells you the size of a serving and how many servings you are allowed or required to eat each day.

Extra Virgin Olive Oil

Olive oil is the main fat in a traditional Mediterranean diet, and the people living in that region tend to consume generous amounts of it daily. Several researchers have studied the role that olive oil plays in the overall health of Mediterranean people. Among other things, their findings demonstrated that this high-fat diet not only did not increase risk but was related to a decrease in cancer in general,[1] specifically a decreased risk of breast cancer in Italy,[2] Spain,[3] and Greece,[4] countries where diets include high quantities of olive oil. The research further showed that the relationship between olive oil and breast cancer seems to be graded; that is, the more olive oil a person consumes, the lower the risk of breast cancer.

Extra virgin olive oil is the natural juice of olives. It is obtained from the first pressing of the fruit—extracted without heat (cold pressed) and without chemicals. It is just olive juice—well, "just" is unfair considering what the components of that juice are and what they can do.

> "I like the way I feel on this diet. I think the olive oil has been good for my body, for my skin and hair."
> —Carol

Ready? Extra virgin olive oil is rich in **antioxidants,** with the highest quantity of any oil of that powerhouse of antioxidants, vitamin E. And extra virgin olive oil is higher in **monounsaturated fat** than any other oil. Why should we care? Because monounsaturated fat does not oxidize in the body, and uncontrolled **oxidation** increases cancer risk. **Polyunsaturated fat,** on the other hand, does oxidize, and extra virgin olive oil has very low levels of it.

Canola oil is also high in monounsaturated fats—a good thing—but no literature supports canola oil contributing to a person's health, except that it is lower in polyunsaturated fats than other vegetable oils.*

---

*The healthy oil claims of canola are in part based on the presence of omega-3 fatty acids. The omega-3's in canola, however, are shorter-chain fatty acids, which our bodies do not

Unlike olive oil, canola oil will not improve such breast cancer biomarkers as blood levels of insulin and glucose, inflammation, and oxidation. Canola oil is simply a neutral dietary fat that is considered healthy because it is high in monounsaturated fat—but olive oil is higher.

As if all this weren't enough to sing the olive's praises, olive oil, unlike canola oil, also contains a number of risk-reducing phytonutrients:

- **Oleocanthal,** a compound that can decrease inflammation in the same way as ibuprofen, which is relevant because nonsteroidal anti-inflammatory drugs have been related to both a decreased risk of breast cancer[5] and its recurrence.[6]
- **Squalene,** which has been shown to be a tumor inhibitor.[7]
- **Lignans,** which have been shown to inhibit the growth of breast cancer cells.[8] Their intake has been related to a decreased breast cancer risk in postmenopausal women from 15 percent[9] to more than 30 percent.[10]

I've saved the best benefit for last. Olive oil increases insulin sensitivity, which might sound like a bad thing but is precisely what you want. Insulin is a hormone produced by your body's pancreas; its job is to lower blood glucose by encouraging it to enter the body's cells, where it can be used for energy or stored for future use. Increased insulin sensitivity means that this hormone works more efficiently, so you don't have to produce as much. In cases where insulin cannot do its job, called **insulin resistance,** the pancreas produces more and more insulin. You may be able to produce enough insulin to keep the blood glucose level in a normal range. (If not, you will be classified as a type 2 diabetic.) However, higher-than-normal blood levels of insulin have been related to breast cancer risk.[11] Losing weight and consuming olive oil both have the ability to decrease insulin levels—ergo, losing weight on a diet that includes olive oil packs a double wallop.

---

use, as opposed to long-chained ones, which are needed to make hormone-like compounds that control such bodily functions as inflammation, blood pressure, and blood coagulation. Because our bodies cannot efficiently convert shorter-chain compounds into long-chain ones, the omega-3's in canola would not contribute to the health benefits associated with omega-3 fatty acids.

Okay. I fibbed. Insulin is not the best part. These two are: Fat makes food taste better and helps decrease hunger between meals. One of the reasons dieters have such trouble adhering to low-fat diets is that they don't feel satisfied with their meals so they eat more, looking for something gratifying, or they finish the diet meal and binge on something between meals. When olive oil is consumed, your meal has flavor and satisfies you. Just wait until you taste those vegetables roasted or braised in extra virgin olive oil. As for decreasing appetite between meals, fat slows digestion, thereby staving off hunger for longer periods of time.

If all of the benefits of olive oil don't have you bathing in the stuff, let me answer questions I am frequently asked.

Does the olive oil have to be extra virgin? Yes. EVOO, as it is affectionately called by those who know and love it, is the least processed olive oil (closest to its natural state), so it has the greatest health benefits.

Does cooking with EVOO destroy its benefits? Nope. That is a myth. Trust me on this. Under normal cooking conditions, the majority of the oil's components are retained. Even when EVOO is subjected to extreme temperature for long periods of time, more than twenty hours, some phytonutrients remain. (Only an observation, but if you're cooking anything at an extreme temperature for more than twenty hours, you'll have bigger problems than the breakdown of phytonutrients! Just saying.)

Does all olive oil taste very strong? Extra virgin olive oil should taste like olives. Other than that, the flavor differs widely depending on a number of factors, such as the variety of the olives, where they were grown, and their ripeness when picked. A useful book for everyone, especially those who are new to extra virgin olive oil, is Deborah Krasner's *The Flavors of Olive Oil: A Tasting Guide and Cookbook.* Krasner suggests that EVOO newbies hold a tasting party. Ask a number of friends to each bring a small bottle of oil and taste each one to determine what suits your taste. Many specialty stores regularly have olive oil tastings.

Isn't extra virgin olive oil expensive? Compared to other vegetable oils, yes. But you're worth it! And if you consider the cost per tablespoon rather than the cost of the entire bottle, EVOO is reasonable, especially when you factor in all it does for you. A 17-ounce bottle has 32 tablespoons, and 1 liter has 64. That means that you are paying less than 25 cents per tablespoon of medium-priced extra virgin olive oil. High-end oils will cost more, and those are the ones I recommend that you use raw (that is, add to a finished soup or for dipping bread), so you can really savor the flavor.

Extra virgin olive oil is the main fat in the PBOO diet and should amount to at least three servings (a total of 3 tablespoons) of your daily allowance of four to five fat servings. I encourage you to include a healthy fat at every meal because, as noted, it satisfies taste and keeps hunger at bay.

Other than olive oil, healthy fat choices are nuts, peanut butter, olives, and avocado. I am a big fan of nuts, especially at breakfast, where a few tablespoons add crunch and character to breakfast cereals. Although I am not aware of any studies that examined the relationship of nuts to breast cancer, they have been related to a decrease in female heart disease. They contain phytonutrients, protein, fiber, and trace minerals such as magnesium and zinc, which are difficult to get into a diet from other sources. All varieties of nuts—almonds, hazelnuts, pecans, pine nuts, pistachios, walnuts, and peanuts, which are actually a legume— are healthy and allowed, but limit or avoid those that are oiled or coated with sugar or salt because it is easy to overeat them.

Aren't nuts high in calories? Sure—about 180 calories per ounce. However, studies have shown an inverse relationship between nut consumption and body weight.[12] A twenty-eight month study in Spain found that people who ate nuts two or more times a week had a significantly lower risk of weight gain, 31 percent, than those who never or almost never ate them.[13] Go figure!

Starch

The PBOO diet recommends that you consume six or seven starches (cereals, grains, beans, pasta, bread) every day and strongly recommends—pleads with you—to make those starches whole grain. Grains are considered whole when they maintain the entire seed or kernel of the plant. The entire seed has three layers: the outer layer (bran), which is loaded with fiber, some B vitamins, minerals, and phytonutrients; the inner embryo (germ), which also contains some B vitamins, plus vitamin E, trace minerals, and phytonutrients; and the starchy middle (endosperm), which is made up of carbohydrates, some protein, and lesser amounts of some B vitamins. It was generous of Mother Nature to provide all that goodness in each tiny grain, so it must be a real slap in her face when we throw out the best part.

When grains are refined, the milling process removes the bran and germ, leaving only the endosperm, the least nutritious part. For some, the resulting product has a more appealing appearance (such as white flour, white bread, white rice), but to get that lack of color, 25 to 90 percent of Mother Nature's gifts are tossed in the trash. Then—something that makes no sense to me at all—companies enrich the stripped grain by adding some, but not all, of the B vitamins and iron that they just removed. And here's the real gobsmacker: So-called enriching may prove to be nearly worthless because research suggests that the beneficial effects of a food are more than the sum of its nutrients. Enriching our foods with nutrients is in no way as advantageous as consuming those nutrients in their original state.

So what are the whole grains? Whole wheat, brown rice, wild rice, oatmeal, whole oats, whole rye, barley, bulgur, and—surprise—popcorn contain the entire seed. Identifying the good stuff should be easy, but it's not always. Breads are particularly troublesome (see the "Choosing Bread Products" section, page 80). Food producers are wise to the growing awareness of the benefits of whole-grain products, so they play a game with us: Confuse the Consumer. "Multigrain," "seven grain," "twelve grain," "stone-ground," "100% wheat," "wheat flour," "cracked wheat," and even "bran" are not necessarily whole-grain products. "Wheat flour" and "100% wheat" are merely trick-the-consumer names for plain old white flour. Oh those jokesters! And even when a product has multi, seven, twelve, or even twenty grains, those grains could be refined grains. You can't even tell by the color because some companies add brown coloring to their products so that they will look like whole-wheat products. To tell whether a product is indeed whole grain, make sure it says it. The first word on the ingredient list should be "whole wheat" or "whole grain."

If your six to seven starch allowances per day are whole-grain starches, you can expect a lot in return. You will be getting a lot of fiber, which is important in all diets but especially for breast cancer because higher fiber can lower blood estrogen levels. Meanwhile, you lower your

triglycerides, lower your blood pressure, improve insulin sensitivity, keep blood levels of glucose in check, ease constipation, and—now, this is really interesting—manage your weight.

Consumption of whole-grain products has been related to both lower body weight and less weight gain.[14] Compared to women who do not consume whole-grain products, women who consume one serving or more per day had a significantly lower body mass index (BMI) and waist circumference.[15] Amazing! Exactly how this happens is a matter of conjecture. Some researchers credit the high-fiber content of whole grains. High-fiber foods take longer to digest, so you feel full longer and are less likely to overeat. I think whole-grain products can lead to lower body weight because they generally have more flavor than refined products. We look for taste when we eat, and if the food delivers it, we will be satisfied with a smaller amount.

> "On the low-fat diet, I was always hungry no matter how much I ate.
> It just did not fill me. On the PBOO diet, I was much more satisfied
> with what I was eating."
> —Mimi

Vegetables

The NCI recommends that we eat at least five servings of fruit and vegetables daily. But lumping fruits and vegetables in the same category is not a good plan when weight loss is the goal. Fruits are higher in calories than vegetables, so a fruit-heavy diet will contain more calories with less food than a diet weighted with vegetables. (See the upcoming "Fruits" section). Furthermore, the literature is more supportive of vegetables being protective for cancer than fruit. It may be that Americans do not eat enough of the healthier fruits to see the same level of protection, but why chance it?

I recommend that you eat at least four servings of vegetables per day, preferably more. A serving size is small—approximately 1 cup for leafy vegetables and ½ cup for most other vegetables. Note that in the PBOO diet, potatoes and plantains are considered starches, not vegetables, and must be counted as such.

Vegetable servings are unlimited. Eat as large a variety of vegetables as possible. Different vegetables supply different nutrients, and you want to cover all bases. Certain vegetables, outlined next, seem to provide better cancer protection, so be sure to include them in your diet as often as possible.

Deep-colored vegetables such as broccoli, carrots, peppers, spinach, beets, tomatoes, pumpkin, and winter squash are particularly high in carotenoids, which have been related to a decrease in breast cancer risk in numerous studies. But—and this is a critical but—you must consume these hardworking pigments with fat to absorb the health benefits. The greatest absorption comes from cooking the vegetable in fat rather than simply adding fat to the meal. Although any fat enhances carotenoid absorption, we already know that olive oil has more going for it than other fats. A study comparing tomatoes cooked in extra virgin olive oil to those cooked in sunflower oil showed that only with the olive oil was there an increase in the antioxidant activity in the blood.[16] The study concentrated on the lycopene in tomatoes, but most likely the same increase in antioxidant action would occur when any vegetable is cooked in EVOO.

Speaking of tomatoes, their deep red color is caused by the concentration of a potent antioxidant, lycopene, linked in many studies to a reduction in cancer risk. Bring on the **lycopene**! And add heat because our bodies absorb lycopene best when the tomatoes are cooked, concentrated, or processed, such as in tomato sauces, tomato paste, and canned tomatoes.

Cruciferous vegetables (broccoli, broccoli rabe, Brussels sprouts, cabbage, and kale) contain the phytonutrients group called **glucosinolates,** which have been shown to stop the spread of cancer and activate enzymes that halt the making of carcinogens. Glucosinolates also alter estrogen metabolism by increasing the metabolism of estrogen to weaker forms that do not promote breast cancer. Here's another but that should be noted: Glucosinolates are water-soluble, so if you boil or steam cruciferous vegetables, you would gain more health benefits by drinking the water than by eating the nutrient-nude vegetables. The healthiest method? You guessed it—cook them in olive oil.

Green leafy vegetables such as spinach and kale are rich in folate, a B vitamin that is needed for the health of your DNA. Increased cancer risk is related to both low intakes and low blood levels of folate.

Salicylic acid, the same anti-inflammatory agent that is an active component of aspirin and has been related to decreased breast cancer risk, is found in broccoli, cucumbers, okra, spinach, sweet potatoes, processed tomatoes, green peppers, radishes, and zucchini. See why you need a variety of vegetables?

Garlic and onions are alliums, which are bulbous plants related to the lily family, and are related to a decreased risk of breast cancer in at least one study.[17] Other studies have found limited or no evidence linking alliums to reduced risk. The few studies that have been made suffer from limitations. For example, questions exist about the accuracy of reporting the amounts and frequency of consumption. Garlic studies are further compromised by the inability to compare data from different garlic sources (fresh, dried, and pill supplements). These limitations make a scientific conclusion about the cancer prevention qualities of garlic and onions extremely difficult, so we can only speculate. Alliums do have compounds known to be protective against cancer and heart disease. For example, they contain sulfur compounds, such as the quercetin found in red onions, which studies have suggested will cause the death of certain cancer cells, including breast cancer cells. Also, garlic and onions constitute a large part of the Mediterranean menu and are most likely a synergistic component of the health benefits of the diet, so I recommend that you include them in your diet—cooked in olive oil, of course.

Suffering from the same limitations as garlic and onions are other vegetables, herbs, and spices, including leeks, scallions (green onions), fresh ginger (gingerroot), chives, oregano, and turmeric. Perhaps you've heard or read about them being cancer protective. Although no direct evidence exists that any of these can decrease the risk of breast cancer, they won't harm you. Plus, they don't add calories and they do add flavor. So go for it!

In a perfect vegetable world, all the produce you eat is from your own garden or purchased from a local grower: fresh, seasonal, and free from pesticides. But that is not always possible. Nancy squirms and hyperventilates when I mention this, but the nutritional values are equal in fresh, frozen, and canned vegetables. In some cases, frozen or canned vegetables may be higher because they are picked and preserved at the height of their ripeness, when their nutrients, especially carotenoids (which account for the deeper color), are most plentiful. Hence, all that goodness

is trapped inside.* Vegetables in your local market, unless they are lo-
cally grown, most likely are days or weeks away from the time they were
picked, so their nutrients are waning. Many vegetables are harvested
before they are fully ripe so they will not rot during their journey to the
market. These poor jet-lagged creatures never realized—and never will
realize—their full nutrient potential. Remember lycopene, that powerhouse antioxidant in tomatoes? Tomatoes ripened on the vine have significantly higher amounts of lycopene than those ripened on your
windowsill.

I encourage people who enjoy planting their own gardens and seeking
out farmer's markets and local growers to keep doing so. Those who
would rather eat dirt than dig in it or brush it off freshly picked vegetables can use frozen and canned vegetables, which provide the same protective benefits as fresh vegetables.

The Minor League

Extra virgin olive oil, starch, and vegetables are daily food requirements.
Other foods are acceptable additions to a good diet but are not required.

Fruits

Fruits are healthy choices, but they are limited to three servings per day.
(See Chart 2, "The Diet Grid," on page 47 for serving sizes because they
vary according to the fruit.) Daily allowances are restricted not because
fruits are not good for you but because most contain enough carbohydrates to make them high in calories. This surprises many dieters,
who think they will lose weight by adding lots of fruit to their diet. But if
you eat one medium pear, one cup of sliced bananas, and a medium ap-

---

*As of this writing, concerns have been raised about the dangers of bisphenol A (BPA),
which is used to line some food cans, primarily infant formulas, soups, and ravioli.
Presently, there are no government guidelines for safe levels. Some animal studies have
shown that exposure in utero and possibly early childhood can lead to later breast or
prostate cancer, but the research is limited and inconsistent. The results of a Harvard Center for Risk analysis, published in 2006, concluded that the existing evidence "does not support the hypothesis that low oral doses of BPA adversely affect human reproduction and
developmental health." BPA is a concern, but how much is not certain. As the damage
seems to be early exposure, limiting or avoiding canned food in pregnancy could help decrease cancer risk for offspring—continue to follow the findings.

ple a day, you have consumed a whopping 405 calories! And these three fruits aren't as powerful as others when it comes to breast cancer prevention.

The deeper the color of the fruit, the better it is for you because the pigment indicates an abundance of carotenoids. So red grapes, berries, apricots, peaches, cantaloupe, and raisins are high in these cancer-fighting phytonutrients; pears, bananas, and apples are not. Because carotenoids are best absorbed when cooked in fat, you probably think I'm going to suggest that you cook them in extra virgin olive oil. Why not? Raisins are delicious sautéed in olive oil with pine nuts, garlic, and spinach or broccoli. Italians often add sautéed raisins to pasta dishes and make delicious cantaloupe chutney with olive oil. Red grapes and dark cherries add great flavor when halved and sautéed in olive oil with chunks of winter squash or pumpkin, with or without chicken. Squash ravioli with cranberries bring raves. Creative cooks won't feel limited cooking fruits in oil, but for those who think of the kitchen as the room with the big white things, remember that if the meal includes fat, you will absorb some of the carotenoids. Adding nuts to the blueberries on your morning cereal and eating a piece of fruit with a meal that contains fat will promote absorption.

Fruits contain other phytonutrients in addition to carotenoids that function to prevent cancer. Antioxidants are found in apricots, berries, and citrus fruits. A high percentage of salicylic acid, the anti-inflammatory agent, is in apricots, all berries, grapes, raisins, cantaloupe, cherries, oranges, pineapple, plums, and prunes.

As in the ideal vegetable world, in fruit utopia, each piece of fruit goes from the organic garden to your mouth and is ripe enough that you have to stand over the sink to eat it. Sigh! But (sit still, Nancy) canned, frozen, and dried fruits contain the same nutrients. However, although fruit juices are as nutritious as the fruit itself, studies show that our brains do not seem to register the calories we drink in the same way as the calories we chew. Consequently, people who drink their calories are far more likely to gain weight. So stick with whole fruits rather than juice.

Dairy and Eggs

Dairy products—milk, cheese, and yogurt—are limited to two servings per day. What? Only two? Dairy products are not essential to a healthy diet, so even those two servings are optional. "Don't I need calcium?" you ask. I'm asked that question a lot. Calcium is important for bone strength, but it is not the only contributor, nor is dairy the only source of calcium. A number of long-term studies have made us question the value of the large quantities of calcium that are currently recommended.* When researchers combined the data from several of those studies, they concluded that there was no relationship between high calcium intake and lower bone fracture risk.[18]

Relevant to the amount of calcium you need is the amount of meat you eat. When our bodies digest meat protein, we release acids into our bloodstreams. To buffer, or neutralize, those acids, our bodies draw calcium from our bones, potentially leading to bone weakness. A few servings of meat a week most likely won't decrease bone strength, but a lifetime of daily meat intake could. A twelve-year Nurses' Health Study found that women who ate more than 3 ounces of meat a day were 20 percent more likely to break a bone than women who consumed less than 2 ounces a day.[19]

For the prevention of osteopenia (decreased bone mass), calcium intake is the usual focus, but a number of other micronutrients are also involved. Vitamin K, found in dark green, leafy vegetables, has consistently been found to improve bone density by stimulating the enzyme that deposits calcium in bone. Eating just one or more servings a day of broccoli, dark green lettuce, kale, or collards is sufficient to meet the recommended amount of 90 micrograms of vitamin K a day for women. Diets high in potassium-rich fruits and vegetables buffer the acid; that is, they reduce the amount of calcium drawn from bones to neutralize acids. Our bodies use vitamin C, found in citrus, peppers, tomatoes, and broccoli, for collagen formation. A high intake of vitamin C equals higher bone mass; a low intake, faster bone loss.

As for nondairy sources of calcium, look to beans, legumes, and dark green, leafy vegetables, such as kale, collard greens, and broccoli. (Spinach and chard also have calcium, but they contain oxalic acid, which changes the calcium's chemical structure, making it unavailable to us. Oh the wonders of how our bodies work!) If you follow the

---

*The National Academy of Sciences recommends 1000 milligrams a day for those age nineteen to fifty, 1200 milligrams a day for those fifty and over, and 1000 milligrams a day for pregnant and lactating adult women.

allowances in the PBOO diet, you will get at least the minimum requirements for calcium even without using dairy.

So what about the vitamin D added to dairy? Vitamin D is more important to bone health than calcium. It is found in dairy products, made
by our skin when we are exposed to sunlight, and available in supplements. The current Recommended Dietary Allowance (RDA) is 400 IU
per day, but researchers who evaluate our need feel that it should be
more than 1000 IU. Those who have limited exposure to direct sunlight,
have darker complexions, or spend winter months in northern parts of
the country may well need 2000 IU. To consume that quantity in dairy,
you better plan on buying the cow. You are best off with supplements,
which work well in this case.

Another question I am often asked is whether milk consumption can
increase breast cancer risk. Contrary to what you may have read or
heard, the last twenty years of comprehensive studies have concluded
that the consumption of milk and dairy products does not increase
breast cancer risk.[20] Some studies found that the consumption of low-fat
dairy products by premenopausal women was related to a decreased
risk. No relationship was found in postmenopausal women.[21]

Some of you may be concerned about the hormones in milk, but no
link has been found between the hormones in milk and breast cancer.
Someday, there may be evidence that hormones in milk are related to
breast cancer, but because hormones have been in use only since 1993,
no substantial, long-term studies exist.

Now that I have, I hope, eliminated your concern about calcium, let's
get back to a general discussion of dairy. You want the majority of the
calories you consume to start improving biomarkers. Dairy products,
even low-fat dairy, can eat up a number of calories without contributing
the same health benefits as foods from the "Big Three"—starch, vegetables, and healthy fats. So dairy, including eggs (which technically aren't
dairy), is limited to two allowances per day.

Milk. Choose nonfat, 1%, or nonfat evaporated milk. The serving sizes
are in Chart 2, "The Diet Grid," on page 47. As for soy milk, the jury is, if
not still out, at least walking in slowly. Soy contains isoflavones, which
are antiestrogenic pytoestrogens; that is, isoflavones compete with estrogen for receptors, which would indicate that they have anticancer
properties. Yet, while some studies show that isoflavones are protective

for breast cancer, others suggest that they can increase risk. Favorable results come primarily from studies in Asian countries, where women maintain a lifelong diet that includes soy products in their natural state—tofu, edamame, and unsweetened soy milk. Studies of Western populations of women who consume soy products are less encouraging. Evidence indicates that if a woman consumes soy products when she is pregnant with a female fetus, that child will develop breast buds that are more resistant to estrogen-positive breast cancer. However, that same protection has not been shown in females who add soy to their diets in adulthood. Bottom line? If you want soy milk, choose an unsweetened, unflavored variety and consume no more than the daily dairy allowance. Stay away from soy supplements because they have not been proven beneficial and may do harm. Also, if you are taking tamoxifen, you should not use soy because it interferes with the drug's absorption.

Cheese. Both full-fat and low-fat (part skimmed) cheeses are allowed, but make sure that each serving does not exceed the calorie limits in Chart 2, on page 47. Choose wisely—a little brie goes a very short way. Perhaps the most difficult part of adding cheese to your diet is controlling the serving size. If you buy a block (brick) of cheese, score it into 1-ounce pieces so that you don't have to guess at the serving size. Keep shredded cheeses on hand because you will find that 1 ounce (¼ cup) of shredded cheese covers more food area than a 1-ounce slice.

Yogurt. Remember when stores carried only two or three choices of yogurt? Today, yogurt decisions require a master's degree in nutrition and a calculator. Before you decide if you want a particular kind, know what's in it. Let's work backwards. Yogurt is milk, curdled by bacteria. Milk has 8 grams of protein per 8 ounces; yogurt should have the same. If the product has more than 8 grams, protein has been added to give a low-fat product more body. This process adds calories but does not add a nutritional advantage. Milk does not contain fiber, so if the yogurt label lists fiber, it has been added, again to no advantage because not enough fiber is added to consider it a health benefit.

As for sweeteners, the latest entries on the yogurt shelf look more like selections from a bakery counter than a dairy aisle. Whether in the form of sugar, high-fructose corn syrup, corn syrup, maple syrup, or the jam that makes the fruit in fruit yogurt, they all add carbohydrates, which translate to calories. A cup of milk has 12 to 17 grams of carbs; if your yogurt has more carbs than that, they are from sweeteners. Your taste for sweeteners will diminish before long, so if you choose yogurt as one of your dairy servings, choose one in a natural form and don't waste those calories.

> After the initial shock of giving up sugar,
> I found that I did not crave or even miss it.
> —*Mary-Ellen*

Eggs. Poor eggs! They are often the subject of character assassination. The truth is, eggs are good for you and may be protective for breast cancer. Some studies have even shown an inverse relationship between egg consumption and breast cancer, especially for later development of the disease in women who consumed eggs as teenagers.[22] Also good to know is that they have never been linked to heart disease and their effect on cholesterol levels is very small. The amount of cholesterol in one egg is equal to the cholesterol in about 7 ounces of beef, and most people who regularly consume beef eat more than 7 ounces.

I like to see eggs in a diet because they are nutritious. (Think about it. They have all the components necessary to support a growing chick.) They are inexpensive and versatile and there is no reason to keep them out of your diet.

Optional Foods in Limited Quantities

Some foods are optional in the PBOO diet. I am not advocating that you include any of these foods in your diet. However, I recognize that some of you may feel that you can't follow a diet that does not include these choices, and I want you to know that it is possible to enjoy the foods you love and still lose weight and improve your health.

Meat, Poultry, and Seafood

Meat, poultry, and seafood are optional on the PBOO diet. If you choose to include them, you are allowed 12 ounces total per week of poultry and seafood, and 6 ounces of beef, pork, and lamb a month in place of (not in addition to) 6 ounces of poultry or seafood. To keep within the calorie limits of the diet (see the "Calories" section in chapter 3), when you include meat in your diet you may have to hold your allowance of foods in other categories to the minimum allocated—do not go below those minimums.

You should restrict your intake of meat, poultry, and seafood for several reasons. First, unlike the Big Three group, which provides phytonutrients that studies show will improve your health, meat, poultry, and seafood do not contain anything that promotes health. Do you recall how olive oil and whole grains have the ability to lower insulin levels, which when high have been positively linked to an increased risk of breast cancer, a recurrence of breast cancer, and death? The amino acids that make up the protein in meat can stimulate insulin secretion. That's no good! Remember how vegetables fight oxidation? Methionine, an amino acid in meat, can be converted in our bodies to a compound called homocysteine, which can increase oxidation. Egad! Consequently, to sacrifice the healthy calories of whole grains, vegetables, or olive oil to consume more meat, poultry, or seafood defeats the purpose of the diet.

Simply adding these foods to the diet in addition to eating the maximum allowances of the other three groups will pack on the pounds. In fact, higher protein intake for women has a stronger relationship to obesity than fat intake does.[23] That's an easy one to explain. A diet that includes a great deal of meat is high in protein, which our bodies do not store. So where does the protein go? It's converted to fat. Those aren't love handles on your sides; they're pork chops!

Am I saying that you have to become a vegetarian? No, although it is a healthy lifestyle and is worth a try. If you miss meat, include it but in its proper place. The risk of cancer, including breast cancer, is related to the ratio of plant products to meat consumption.[24] The more plants and less meat you eat, the lower the cancer threat.

For at least the first two weeks of your diet, I strongly recommend that you do not include any meat, poultry, or seafood in your diet. Abstaining from these three will give you a promising jump start to shedding pounds. When I asked the women in my study to do this, they groaned, thinking they couldn't handle it. Within two days, all of them said that it wasn't so bad after all. Another advantage to eliminating meat, poultry, and seafood is that you can begin to think of vegetables as something more than boiled, baked, or steamed accompaniments. Try combining several different vegetables, add herbs and spices, and unite them with beans and rice, and use them in pasta sauces.

Studies connecting red meat, processed meat, and grilled, barbecued, and high temperature pan-fried meat to breast cancer have been inconsistent. Over the years, a number of researchers have related all of these to increased risk. Contradicting those findings, a recent NIH-AARP diet and health study found no relationship between meat intake (regardless of preparation methods) and breast cancer.[25]* Interesting. But look at Table 1 of the study and you find that the women with the lowest intake of red meat had an average BMI of 25.3 compared to 28.1 for the highest consumers. What is the significance of that 3-point difference? If you are 5 feet tall and in the highest consumption bracket, those 3 points mean 14 extra pounds; a 5-foot, 8-inch woman would add 19.5 pounds. No studies deny the relationship of excess weight to breast cancer. I am 5 feet tall, and 14 additional pounds would be not just excess weight but a health catastrophe waiting to happen.

Grilling, barbecuing, and high, dry heat create compounds called heterocyclic amines (HCA), which in animal studies have induced mammary tumors.[26] Similarly, the Iowa Women's Health Study linked well-cooked meat to increased incidents of breast cancer.[27] The NIH-AARP study did not support this relationship. Am I ready to say, "Bring on the grilled t-bone?" Absolutely not. I don't think that consuming large quantities of red meat will ever be considered healthy, so keep the servings small and infrequent.

Alcohol

Does consuming alcohol increase the risk of breast cancer? Many people will tell you that it does—and others will say the studies are inconsistent. Some studies show that the risk is related only to estrogen-positive breast cancer; others, to estrogen-negative breast cancer. A true relationship would be consistent for cancer type. And then there is the question of amounts.† Some studies have found an increased risk with more

---

*The study assessed only postmenopausal women, and the authors did not rule out the possibility that eating meat and exposure to meat mutagens at a younger age—particularly during adolescence when the breasts are developing—may increase the risk of breast cancer.

†For study purposes, a drink has 12 to 15 grams of alcohol, which equates to 1 ounce of spirits, 12 ounces of beer, and 5 ounces of wine.

than two drinks a day, but other studies have suggested the risk exists with fewer drinks. A Swedish study of close to forty thousand alcoholic women concluded, "It is therefore apparent, contrary to expectation, that alcoholism does not increase breast-cancer risk in proportion to presumed ethanol intake."[28] Worth noting is that even in studies that found increased risk, the threat was small—nowhere near the risk related to being overweight.

I am not suggesting that you should begin making daily runs to the liquor store or start hanging out in bars. I am not condoning excessive drinking. But I do not think there is sufficient evidence to suggest that a glass of alcohol a day can increase the risk of breast cancer. A 2008 study in southern France found that women who drank approximately a glass of wine a day had a lower risk of breast cancer compared to non-drinkers.[29] Another study in Seattle found that women who were moderate drinkers (regularly consumed seven drinks a week) and were diagnosed with breast cancer before they were forty-five had a 30 percent decreased mortality risk from the disease compared to non-drinkers.[30]

Red wine may actually be beneficial because it contains polyphenols, a phytonutrients category that has been shown in test tubes to act as an antioxidant and decrease the proliferation of breast cancer cells.

Exercise: As Important as What You Eat

Some of us love exercise. Many of us hate it. But we all need exercise, so make it part of your new way of life. Exercise will help you lose weight by burning calories and adding muscle—muscle that requires more calories to maintain than fat. Once you lose weight, don't stop exercising because it is critical to maintaining weight loss.[31] Exercise is also related to a decrease in blood glucose, which decreases cancer risk.

I recommend that you exercise a minimum of four times a week. Because you start to use stored fat for energy only after thirty minutes of exercise, try to exercise for at least thirty consecutive minutes, with a goal of one hour. Exercise as long after (or before) eating as possible. Your insulin is lowest at that time, so your body will release fat stored in cells to use as energy. Translation: Instead of using the cereal you just ate to supply energy for exercising, insulin will take it from that cheeseburger you ate four months ago that is still sitting on your hips. One more thing about burning fat: it needs oxygen to be broken down, so don't get breathless. You should be able to talk while you are moving. Swearing counts.

In addition to a regular exercise routine, try to find ways to move more every day: Walk to anything within a mile of where you are starting, use stairs for as many floors as you can, get off the bus a stop before your destination, and choose a parking space as far from the entrance as possible.

As you can see from reading this chapter, the PBOO diet concentrates much more on what you should eat than on what you should not. When you consume the required foods in the given amounts, you will find that your cravings for the foods you should avoid will diminish or disappear. And then . . . well, read what the women themselves had to say.

I felt better on this diet than frankly I had ever felt on any other diet or
nondiet food plan. My weight has remained constant since completing the study
four years ago. It is quite ironic that after a lifetime of up and down
weight swings it took breast cancer to help me feel really well.
—*Christine*

I loved the fresh vegetables that I learned to roast in the oven with olive oil.
I learned to make a killer barley salad with chopped fresh cucumber,
tomatoes, and peppers, seasoned with capers and black olives.
—*Mary-Ellen*

The plant-based diet was different than other diets I had experienced in
the past. It took some time to adjust to eating more vegetables and adding
extra virgin olive oil as well as proteins other than meat. I had never tried beans
or legumes, other than maybe chickpeas on my salad. Much to my surprise,
I started to really enjoy the wonderful flavors of olive oil with herbs,
vegetables, and beans. Even my family started to enjoy several of the
recipes provided by the program.
—*Cheryl*

The plant-based olive oil diet enabled me to lose weight (about 15 pounds)
without feeling hungry or depriving myself of foods that I love. I found the
recipes easy to follow, tasty, and made with ingredients that I was familiar with
(cooking is not a hobby of mine). Most of the ingredients could be stocked in
the kitchen so that I could always put together a good meal without making
numerous trips to the grocery store. My husband also enjoyed the meals,
and we both had greatly improved cholesterol readings.
—*Lois*

I really became a cook as a result of the plant-based diet.
I lost weight while really enjoying food.
—*Mary Ann*

What I love about the plant-based diet is that I can have a great deal
of food . . . it's not bad stuff, and I can have a wonderful variety. I don't have
to be so-o-o precise about the amount of spinach I'm using. . . .
Plus, my food looks so beautiful because it's very colorful.
—*Renee*

My entire family can follow the diet. The six grandchildren—ages five to
thirteen—look forward to Gramma's cooking when they visit. My husband,
Mike, has a small organic vegetable garden in the summer that he harvests
through the fall. We are never at a loss for vegetables.
He lost a lot of weight following this diet.
—*Fran*

A New Way of Life

Obstacles are those frightful things you see when you take your eyes off your goal.

—Henry Ford

Getting Started

Noted educator Laurence J. Peter, best known for his book *The Peter Principle,* said it well: "If you don't know where you are going, you will probably end up somewhere else." Start your new life by setting a weight loss goal. Purchase a dependable scale and weigh yourself at least once a week, noting your progress on paper. You should plan on losing no more than one or two pounds a week or an average of about four pounds a month. Fat takes longer to lose than muscle, so if you drop pounds too quickly, muscle is what you'll be shedding. And because it takes more calories to maintain muscle than fat, the more muscle you have, the more calories you will burn. Don't be discouraged if you don't see a huge weight loss immediately. Remember that your diet is also bettering your health, and improvements come quickly in that area.

Stocking the PBOO Pantry

Healthy eating begins with healthy grocery shopping. But before you start filling the shopping cart, show your kitchen that you mean business and get rid of all problem foods—chips, baked goods (including nonfat ones), candy, ice cream—anything that is your go-to food when your resolve wanes. Replace those foods with permitted foods.

I threw out all the junky food as recommended. It was very liberating. I knew
I would be successful after that ceremony. Then I went grocery shopping.
—*Fran*

Be sure to have a generous supply of items that you can turn into a
PBOO meal quickly; that way, you will be less likely to fill up with unhealthy calorie choices. Chart 1, "The PBOO Pantry," lists some of the
items you should have on hand in the cupboard, refrigerator, and
freezer. These items are "the right stuff" for healthy—and in many cases
fast—meals. More than likely you have a favorite food, condiment, or
relish that is not listed. If it does not contain trans fats, is not a cured
meat, and does not contain long, unpronounceable ingredients, it is
probably neutral and you can use it if it is within the calorie range of
that food category.

Set aside a time and place to eat. Make dining a ritual whether you
are eating alone or with your family. So much of the weight we gain
comes from spontaneously grabbing fast-food items because we are
hungry and have no plan of what to eat when. For the most efficient
weight loss, eat only three meals a day and do not snack in between—
not even "healthy" snacks. During that in-between time, your insulin is
at its lowest and will draw energy from stored fat. If you must have the
occasional snack, choose one from the allowed list and make sure you
count it.

Cooking your own food on a regular basis will help establish a pattern
of choosing healthy food, so try not to eat out in restaurants for at least
the first two weeks. After you have a well-established, healthy eating
program, you will be less likely to be broadsided by a cheeseburger or
tackled by a piece of chocolate cake. When the pounds start to drop,
your resolve to keep going will decrease temptation. Bring your lunch or
dinner with you if you have to eat away from home—at work and even at
friends' houses. A good friend will understand your need to diet. A really
good friend will serve everyone the same thing you are eating.

I never snacked between meals. I carried my own food with me to parties,
picnics, and lunch with friends. It was easy to do by simply telling them
that I was in an important, life-saving, hospital diet study.
—*Mary-Ellen*

Chart 1. The PBOO Pantry

Starches

In the cupboard: high-fiber, low-sugar breakfast cereal, oatmeal—instant and old-fashioned oats, steel-cut oats, variety of whole grains—brown rice, barley, quinoa, orzo, a variety of dried and or canned beans, whole wheat pasta in an assortment of shapes (including lasagna), whole wheat flour.

In the refrigerator: an assortment of whole wheat breadstuffs such as pita, wraps, tortilla, lavash, sandwich rounds, English muffins, sliced bread. (See "Choosing Bread Products," page 47.) These items keep longer when refrigerated.

Vegetables

In the cupboard: canned artichoke hearts, corn, peas, jarred or canned roasted red peppers, salsa, canned tomatoes—whole, diced, crushed—with or without added seasoning.

In the refrigerator vegetable bin: red and yellow onions, garlic, carrots, celery, potatoes (Yukon gold, baking, red), peppers, mushrooms, baby spinach, parsley, ginger, salad greens. Keep opened cans or tubes of tomato paste on a refrigerator shelf; the tubes keep indefinitely but the opened cans have a shelf-life of about a week.

In the freezer: An assortment of vegetables (plain with no sauce added). A good selection would be broccoli, butternut squash, cauliflower, corn, peppers, spinach, and a variety of other greens.

Fruit

In the cupboard: canned apricots, peaches, pineapple, as well as tropical mix, raisins, and other dried fruits. Choose dried fruits with no sugar added.

In the vegetable bin: ripe bananas, seasonal fruit, berries, lemons.

Fat

In the cupboard: extra virgin olive oil, peanut butter (natural or without trans fatty acids), canned olives.

In the refrigerator: loose olives (that is, those you buy by the pound), peanut butter after opening, is recommended.

In the freezer, especially if you buy in quantity: a variety of nuts, such as chopped and whole walnuts, whole and slivered almonds, pecans, and pine nuts.

Dairy

In the refrigerator: large eggs, nonfat or 1% milk, yogurt, a variety of cheeses, shredded cheese for quick meals.

Seafood

In the cupboard: tuna packed in water, canned clams, vacuum-sealed pouches of salmon, tuna, and so on.

Miscellaneous

In the cupboard: canned broth (vegetable and chicken), sea salt, and vinegar; an assortment of dried herbs and spices, such as oregano, thyme, peppercorns, paprika, curry powder, cinnamon, red pepper flakes, and nutmeg.

In the refrigerator after they are opened: maple syrup, soy sauce, capers, mustard.

The Diet Grid and Grid Eating

The core foods and the optional ones with their allowed amounts are listed in Chart 2, "The Diet Grid." (Although the chart is not literally a grid, Nancy began calling it that and the name stuck. Now she refers to following a healthy diet as "grid eating." Works for me). Familiarize yourself with the minimum and maximum allowances so you can plan an entire day of eating. That way, you won't find yourself trying to consume an abundance of starch or fat at one meal because you forgot it earlier in the day, or discover that the toast slathered with five tablespoons of peanut butter that you ate for breakfast used up your daily fat allowance. Making daily meal plans and sticking to them will not only prevent you from overeating or eating "wrong" foods but will also ensure that you consume the necessary amounts of "right" foods to improve those biomarkers. Meal planning requires some effort in the beginning—grocery shopping for allowed foods, checking labels, weighing and measuring for meal preparation. But you will soon learn the basics and find yourself grid eating without thinking about it. Note that when a food is less than ½ a serving size, you do not have to count it.

In addition to weighing yourself, you will want to weigh and measure food for accurate portion sizes. Have on hand dry and wet measuring cups, a set of measuring spoons, and most important, a reliable food scale.* Before long, you should be able to eyeball serving sizes of many foods. One serving of rice, for example, is no bigger than a tennis ball; a serving of chicken is about the size of a deck of cards or a hockey puck. However, don't eyeball the serving size of fats; always measure.

---

*Nancy and I both use Salter electronic scales, with enthusiasm. This two-hundred-year-old company knows scales! The readouts are clear, with ⅛-ounce precision. With the press of a button, the scale converts metric to imperial weight, and vice versa. Some models calculate nutrients. The best feature is the zeroing-out capability. Place an empty bowl on the scale, press a button, and the reading is back to zero. Put an ingredient in the bowl, get the weight, press a button, put another ingredient in the bowl, and you'll get the weight of only that ingredient. Salter scales cost about $30 and upward and usually come with a ten-year guarantee. Nancy has used hers for thirty-seven years without fail.

Chart 2. The Diet Grid: Serving Sizes and Allowances

STARCH*

Six to seven servings per day
80 to100 calories per serving

> For maximum benefit, choose whole-grain starches (see page 26).

Beans: (All types of plain, unflavored beans)
> 1 ounce dry, ½ cup cooked or canned

Breadstuffs: (Serving allowances below may vary; see page 80 for information on choosing breads)
> Bagel, ½ small (1 ounce)
> Bread, 1 slice (1 ounce)
> English muffin, ½ to 1
> Pita bread, 1, 6½ inch
> Tortilla, whole wheat: ½ to 1, 6- to 9-inch
> Tortilla, corn, 1, 6-inch
> Wrap, ½ to 1, 7-inch

Cereal: Avoid cereals loaded with sugar. Check labels for serving size and calories; a serving is 110 calories or less.
> Cold: ½ to ¾ cup (1 ounce)
> Hot: ¼ cup dry; ½ cup cooked

Grains:
> Couscous, 1 ounce dry, ½ cup cooked
> Barley, 1 ounce dry, ½ cup cooked
> Bulgur, 1 ounce dry, ½ cup cooked
> Rice, 1 ounce dry, ½ cup cooked
> Orzo, 1 ounce dry, ½ cup cooked

Pasta: 1 ounce dry

Plantain: Although a fruit, plantain counts as a starch for diet purposes.
> 2 ounces raw, ½ cup cooked and mashed

Potatoes: All varieties of white, sweet, and yams are vegetables that count as starches.
> 3 ounces raw, ½ cup cooked

Popcorn, air-popped, 3 cups

VEGETABLES*

At least four servings per day
10 to 30 calories per serving

> Vegetable allowances are unlimited. The serving sizes are given to ensure that you eat the minimum.

> With the exception of leafy vegetables, approximately ½ cup of raw or frozen vegetables and slightly less of cooked vegetables equals one serving.

Leafy greens: 1 cup raw; ⅓ cup cooked or frozen

Salsa: ¼ cup (with no oil added)

(continues)

Chart 2 *(continued)*

FAT*

Four to five servings per day
100 to 120 calories per serving

Fats help to satisfy hunger and curb appetite between meals, so try to include one of the healthy fat servings below at every meal. At least three servings per day of fat should be from olive oil.

Extra virgin olive oil: 1 tablespoon

Nuts, all varieties: 2 tablespoons (almonds, pecans, walnuts, cashews, and so on; avoid nuts that are coated with salt or sugar)

Peanut and other nut butters: 1 tablespoon

Sunflower seeds: 2 tablespoons

Pine nuts (pignoli): 2 tablespoons

Avocado: ¼ (2 ounces trimmed weight)

Olives:
Pitted, 12 large or 20 small (2¼ ounces)
With pits, 12 large or 20 small (4 ounces)

FRUIT

Three servings per day
50 to 70 calories per serving

As with vegetables, deep color indicates a high percentage of carotenoids, so fruits such as red grapes, berries, apricots, peaches, and raisins are good choices.

Fresh fruit: Weights include the skin or peel and seeds and are meant to be used when purchasing the fruit.
Apple, 4 ounces
Apricots, 4 (5 ounces total)
Banana, 4-inch piece (½ cup slices)
Berries, 1 cup
Cantaloupe, ¼ medium (1 cup pieces)
Cherries, 13
Figs, 2 small (3 ounces total)
Grapefruit, ½ medium (1 cup pieces)
Grapes, 17
Kiwi, 1½ (½ cup)
Mango, ½ cup (3 ounces)
Nectarine, 1 (5 ounces)
Orange, 1 (6 ounces)
Papaya, 1 cup (6 ounces)
Peach, 1 (5 ounces)
Pear, ½ (4 ounces)
Pineapple, ¾ cup pieces
Plums, 2 small (5 ounces total)
Tangerines, 1 large (6 ounces)
Watermelon, 1¼ cup

Chart 2 *(continued)*

Canned fruit, in juice or water, and ***applesauce,*** with no sugar added, ½ cup

Dried fruit:
    Apricots, ¾ ounce
    Cherries, 2 tablespoons
    Cranberries (Craisins), 2 tablespoons
    Figs, 1 (¾ ounce)
    Prunes, 3 pieces (1 ounce)
    Raisins, 2 tablespoons

DAIRY

Two servings per day
80 to 110 calories per serving

Milk:
    Nonfat or 1%, 1 cup
    Nonfat evaporated, ½ cup

Eggs: 1 large

Cheese:
    Hard, ¼ cup grated or crumbled, 1 thin slice (1 ounce)
    Cottage cheese, 1% fat: ½ cup (4 ounces)
    Feta, ¼ cup
    Ricotta, part skim, ¼ cup (4 ounces)

Yogurt: Plain, low-fat, or nonfat, ½ cup (4 ounces)

MEAT/POULTRY/FISH

12 ounces per week
30 to 60 calories per ounce

For the best jumpstart to the diet, avoid eating meat, poultry, or fish for the first two weeks.

You are allowed 12 ounces total per week but should restrict red meat to no more than 6 ounces a month.

Some healthy choices follow with calories per ounce.

Fresh fish:
    White: 25
    Oily (Atlantic Salmon, Mackerel): 60

Water-packed fish and shellfish in cans or pouches:
    Tuna: 35
    Salmon: 40
    Crab: 30
    Shrimp: 30
    Clams: 40

Poultry: Skinless chicken or turkey: 35
    For bone-in poultry, calculate weight at 65% of total weight

*You must include the minimum amounts from all the starred categories every day.

Meal Planning

If you need help in planning meals, refer to Chart 3, "Weekly Meal Plan," where you will find a week's worth of sample meals. You can follow the plan exactly or make substitutions from the recipes included in Part Two or your own collection, always paying attention to the serving sizes of the ingredients. The allowance guidelines in Chart 2 for breakfast, lunch, and dinner will help you design meal plans. These are guidelines; there will be times when you want to use, for example, only one starch at breakfast and save the extra for lunch or dinner. However, be sure to include at least one serving of fat at every meal; that will help carry you through to the next meal without feeling that you have to annihilate the contents of a vending machine, wrappers and all.

Calories

Calories do count, but you don't have to count them. A 1500-calorie diet is built into the eating plan.* By sticking to the food allowances listed in Chart 2—six to seven starches, four to five servings of fat, for example—your daily intake of calories will wind up being approximately 1500 calories. In case you are calculating how many brownies equal 1500 calories, keep in mind that you don't want just any old 1500 calories; you want the best, the ones from foods associated with improving breast cancer biomarkers. The goal of the diet is to consume enough of the right foods in sufficient amounts to improve your health. For that reason, it is important that you consume at least the minimum amounts of the starred categories (starch, vegetables, and fat).

For neutral foods, that is, those that are not linked to increased risk—such as condiments, relishes, seasoning liquids, and an occasional

---

*The conventional calorie count for dieting women is 1000 to 1200. For years, I have used 1500 calories with research subjects and clinic patients and they lose weight. A 1000- to 1200-calorie diet is hard to maintain and makes it difficult to include all the necessary nutrients. Furthermore, a lower-calorie diet can adversely affect a person's basal metabolic rate (BMR), which is the energy used by the heart, lungs, and other vital organs. Eating too little food will decrease BMR so that the body goes into energy conservation (survival) mode. When BMR slows, weight loss decreases and people feel tired. In a number of cases, people reported reaching a weight loss plateau and had to slightly increase their calories in order to start losing weight again.

snack—count the calories and feel free to add them if your daily allowance permits.

When you automatically think of your daily intake of food in terms of eating what your body needs, and not calorie restriction, you will be well on your way to a healthier you.

Each food category has a range of calorie counts so you will know the serving size and can use it as a guide when grocery shopping. One serving of starch, for example, should be between 80 and 100 calories. So if 1 cup of cereal is 80 calories, one serving is 1 cup; two servings is 2 cups (160 calories). After you become familiar with the allowance "calorie cost" of the foods you choose, you should be able to plan a meal without continually referring to the chart.

Breakfast

Breakfast on a 1500-calorie diet should have about 400 calories. The ideal breakfast includes food from the following categories, in these amounts.

Two servings of whole-grain starch
One to two servings of fat (nuts and nut butters are particularly
    easy to incorporate into breakfast)
One serving of deep-colored fruit
Optional: One or two servings of dairy and any amount of
    vegetables

Lunch

Lunch on a 1500-calorie diet should have 500 to 600 calories. The ideal lunch includes food from the following categories, in these amounts.

Two servings of whole-grain starch or of beans
Two or more servings of vegetables
1 to 2 tablespoons extra virgin olive oil
One to two servings of fruit, preferably deep-colored
Optional: Cheese, egg, nuts (substitute for olive oil), some of your
    meat, poultry, or seafood allowance

Dinner

On a 1500-calorie diet, dinner should have between 500 and 600 calories, depending on the calories consumed at breakfast and lunch.

Three to four servings of whole-grain starch or beans
Two or more servings of vegetables
1 to 2 tablespoons extra virgin olive oil
One serving of fruit, preferably deep-colored
Optional: Cheese, some of your meat, poultry, or seafood
    allowance

GOOD TO KNOW

Weights and Measurements

Grains, cereals, and breadstuffs are often listed in grams on the package. Knowing how to quickly convert grams to the 1-ounce allowance per serving of starch is helpful.

1 ounce equals about 28 grams
2 ounces equals about 55 grams

Chart 3. Weekly Meal Plan

Day 1	Amount	Starch	Vegetables	Fat	Fruit	Dairy
Breakfast						
Dry breakfast cereal	1 oz.	1				
Non-fat milk	½ cup					1
Raisins	2 TBSP				1	
Walnuts	2 TBSP			1		
Lunch						
Simple Vegetable Soup (recipe page 86)	1 serving	2	3	2		
Fruit	2 servings (see Chart 2 for choices)				2	
Dinner						
Penne with Spinach and Beans (recipe page 164)	1 serving	3	4	2		
Daily totals	*Calories: 1430*	6	7	5	3	1

DAY 2	Amount	Starch	Vegetables	Fat	Fruit	Dairy
Breakfast						
Hearty Oatmeal (recipe page 62)	1 serving	2		1	1	¼
Lunch						
Cheese and Broccoli Wrap (recipe page 97)	1 serving	2	2	1		1
Fruit	2 servings (see Chart 2 for choices)				2	
Dinner						
Rice with Corn, Black Beans, and Tomatoes (recipe page 141)	1 serving	2	3	2		
Daily totals	*Calories: 1475*	6	5	4	3	1¼

(continues)

Chart 3 *(continued)*

DAY 3	Amount	Starch	Vegetables	Fat	Fruit	Dairy
Breakfast						
Dry breakfast cereal	½ ounce	0.5				
Non-fat milk	¼ cup					0.5
Whole wheat english muffin	½ muffin	1.0				
Peanut butter	1 TBSP or 16 grams			1		
Banana	½ cup sliced or 6-inch portion				1	
Lunch						
Basic Potato Salad (recipe page 116)	1 serving	3	3	2		
Fruit	2 servings (see Chart 2 for choices)				2	
Dinner						
Vegetable Lo Mein (recipe page 180)	1 serving	2	5	2		
Daily totals	*Calories: 1450*	*6.5*	*8*	*5*	*3*	*0.5*

DAY 4	Amount	Starch	Vegetables	Fat	Fruit	Dairy
Breakfast						
Whole wheat bread	2 slices	2				
Creamy peanut butter	2 TBSP (32 grams)			2		
Banana	½ cup sliced or 4-inch portion				1	
Lunch						
White Bean and Spinach Salad (recipe page 114)	1 serving	2	4	1		
Fruit	2 servings (see Chart 2 for choices)				2	
Dinner						
Potato, Zucchini, and Fresh Tomato Casserole (recipe page 126)	1 serving	3	5	2		1
Daily totals	*Calories: 1495*	*7*	*9*	*5*	*3*	*1*

Chart 3 *(continued)*

DAY 5	Amount	Starch	Vegetables	Fat	Fruit	Dairy
Breakfast						
Breakfast Tortilla (recipe page 75)	1 serving	2	1	1		0.5
Lunch						
Cheese, Pepper, and Onion Chapati (recipe page 96)	1 serving	2	3	1		0.5
Fruit	2 servings (see Chart 2 for choices)				2	
Dinner						
Pasta with Artichokes and Roasted Red Peppers (recipe page 172)	1 serving	2	3	2+		
Daily totals	*Calories: 1490*	*6*	*7*	*4+*	*2*	*1*

DAY 6	Amount	Starch	Vegetables	Fat	Fruit	Dairy
Breakfast						
Breakfast Egg and Vegetable Sandwich (recipe page 77)	1 serving	2	2	1		1
Fruit	1 servings (see Chart 2 for choices)				1	
Lunch						
Grilled Cheese and Spinach (recipe page 98)	1 serving	2	2	2		1
Dinner						
Bulgur with Roasted Green Beans and Tomato (recipe page 148)	1 serving	2	4	2		
Daily totals	*Calories: 1490*	*6*	*8*	*5*	*1*	*2*

(continues)

Chart 3 *(continued)*

DAY 7	Amount	Starch	Vegetables	Fat	Fruit	Dairy
Breakfast						
Low-fat vanilla yogurt	1 cup					1 or 2
Almonds, dry roasted	2 TBSP			1		
Blueberries	1 cup				1	
Lunch						
Tomato-Pepper Pasta Salad (recipe page 116)	1 serving	3	4	2		
Fruit	Half serving (see Chart 2 for choices)				0.5	
Dinner						
Vegetable-Stuffed Baked Potato (recipe page 123)	1 serving	3	3.5	2		
Daily totals	*Calories: 1490*	6	7.5	5	1.5	1 or 2

I felt that 1500 calories provided me with an adequate amount of food. The food was delicious and filling.

—*Mary Ann*

Sometimes when I entertain, I take a holiday from the diet and cook the recipes I love. Then I make sure that every single leftover leaves the house with the guests.

—*Caroline*

I found that my grocery bill went way down. No meat, no sweets: big savings.

—*Mary-Ellen*

Part Two

The Recipes

Just about all the recipes in our book are intentionally and surprisingly simple. Most can be made in less time than it takes to order takeout. Just a few of them require more than a handful of ingredients. (Nancy couldn't resist developing *some* recipes that use a considerable number of ingredients, yet they are still easy to make.) When I developed the recipes for the Komen study, my intention was to make them as easy—and affordable—as possible, so the study participants would find that the diet did not entail hours in the kitchen or a layaway account at the supermarket. Some of the women never varied from the recipes as written; others were more adventurous and began to adapt them to fit their own tastes: They added creative mixtures of herbs and spices, splashes of vinegar, lemon juice, or soy sauce. Garlic-lovers added cloves of it where there were none and increased it by fistfuls when it was in the recipe. (I use garlic with restraint since I don't digest it well.) You can do the same. Nancy added many suggestions for additions that she liked.

I invented what I called my "Breakfast Mix" and brought the recipe to our group. It consists of a mixture of many dried fruits chopped up, with sesame seeds, walnuts, unsweetened coconut, peanuts, a tiny bit of granola, and sometimes some chopped dried ginger to give it a bite. I sprinkle it on my cereal most mornings. It's based on the Chinese theory that one should eat fifty or sixty different foods every day to keep one's *chi* in balance.
—*Mary-Ellen*

About Serving Sizes

I wrote the majority of the recipes for one person since the women in my program did not initially expect to feed anyone but themselves with "diet" food. But as soon as they found how much they liked the dishes, they asked for ways to increase the recipes so their families and friends could also enjoy them. In most cases, increasing the recipe is as simple as multiplying the ingredients by the number of people to be fed. When a recipe might present questions about increasing the size, we have given suggestions in a box called Increasing the Recipe. When you cook for more than one, you will have to pay some attention to your own serving so that you receive the proper proportions of all the ingredients. For example, if you have increased a recipe for pasta with vegetables by four, your portion should be no more than a quarter of the pasta and at least a quarter of the vegetables.

All the recipes are followed by a list that gives calories and food counts. These are always for one serving, not for the entire recipe.

Ingredients

In most cookbooks, ingredients first list the "market size" (that is, what you buy at the market); so a typical listing would be "1 large onion, chopped (1 cup)." But we have reversed that line to read "1 cup chopped onion (about 1 large)" so you can quickly assess the amount of vegetables in terms of the allowances in the Diet Grid (page 47). We have also given you choices of fresh, frozen, or canned vegetables, and homemade tomato sauce or a jarred variety. We just want you to follow the diet whatever your cooking preferences are.

> I decided to have a cooking day once a month. I roast vegetables; cook rice, barley, and quinoa; portion out red and pink and white beans; chop dried fruit and nuts for the breakfast mix. I make portion-size containers of each food and put them in the freezer. That way I can simply pick out a grain, some vegetables, and a bean portion, then defrost them and have a meal in a few minutes.
>
> —*Mary-Ellen*

Breakfast

A common mistake many dieters make is skipping breakfast. The irony is that those who eat in the morning control their weight better, especially long term, than those who bolt out the door on an empty stomach. That's an established fact. Studies demonstrate that there is an association between not eating breakfast and a higher BMI; one study found that people who did not regularly eat breakfast were over *four times* more likely to be obese.[1] Studies also show that people who eat breakfast are much more productive in their morning work than those who don't.

So, eat breakfast for sure—just don't do it right away. Delay it as long as possible. (That doesn't mean not eating until lunchtime and calling it breakfast). I suggest delaying breakfast because when you wake up in the morning, your insulin is as low as it will be for the day and your body will use stored fat for energy. When insulin is low, the enzyme hormone-sensitive lipase, which works to release fat from fat cells, is turned on; after a meal, when insulin is higher, the enzyme lipoprotein lipase, which is used to take fat in, takes over. For that reason, it is extremely beneficial for weight loss to exercise in the morning before you have eaten.

Patients who tell me they don't eat breakfast usually claim they just don't have time to fix it. Here are a few suggestions for quick, healthful starts to the day:

1. Whole wheat toast with peanut butter (as natural as possible with no trans fats)
2. High fiber, unsweetened cold cereal with milk or yogurt, 2 tablespoons nuts, and 1 serving of fruit

3. Healthy breakfast breads and muffins (pages 65–73). They make quick morning meals if you bake ahead and freeze (see box on page 67).
4. Oven-Roasted Granola (page 63) and yogurt.
5. Breakfast Egg and Vegetable Sandwich (page 77). Takes a wee bit longer but is still a quicker and unquestionably healthier breakfast than you can get from a fast-food place.

Hearty Oatmeal

Microwaves aren't for everything, but for oatmeal they work like a charm. Many a rushed morning I've eaten it right out of the Pyrex measuring cup, saving both time and a dirty dish. Of course, you can also make this by stirring the oats into the boiling water on top of the stove. Either way, the recipe can be successfully halved, so if you want toast or half an English muffin with your oatmeal, you'll still have just two starch servings.

½ cup dry, old-fashioned (rolled) oats
1 tablespoon brown sugar or maple syrup
¼ to ½ teaspoon salt
2 tablespoons walnuts
2 tablespoons raisins
¼ cup milk (nonfat or 1%)

Place 1 cup water in a 2-cup Pyrex measuring cup. Stir in the oats, brown sugar, and salt. Microwave on medium for 5 minutes and stir. Stir in the nuts, raisins, and milk.

MAKES I SERVING (ABOUT ¾ CUP)
CALORIES: 350; STARCH: 2 (oatmeal); VEGETABLES: 0; FAT: I (nuts); FRUIT: I (raisins);
DAIRY: less than ¼ (milk)

Oven-Roasted Granola

Olive oil granola! Are you kidding? I hear that all the time, but I've never failed to make a convert out of a taster. This is so good, in fact, that I have to warn the women in my groups to be sure and measure how much they are eating. It is easy to get carried away. A great topping for yogurt.

2 cups old-fashioned (rolled) oats
¼ teaspoon salt
½ cup honey, maple syrup, or a combination
¼ cup extra virgin olive oil
½ teaspoon cinnamon
½ cup slivered almonds
½ cup raisins or dried cranberries

Equipment needed: jelly-roll pan, at least 15½ × 10½ inches

Preheat the oven to 350°F.

Combine the oats and salt in a mixing bowl. Make a well in the center and pour in the honey, olive oil, and cinnamon. Mix thoroughly with a fork. The mixture will be sticky. Add the almonds and combine well.

Transfer the mixture to the jelly roll pan and use a fork to spread evenly. Bake for 15 minutes. Stir the mixture with a fork to separate. Bake for 5 to 8 minutes longer, until golden brown. Keep an eye on it as the granola can burn quickly. Transfer the granola to a cool cookie sheet or clean cutting board and cool. If you allow it to cool on the pan, it will harden and be difficult to remove. When cool, stir in the raisins. Store in an airtight container.

MAKES 12 SERVINGS, ⅓ CUP EACH
CALORIES: 180; STARCH: ⅔ (oats); VEGETABLES: 0; FAT: 1 (⅓ olive oil, ⅔ nuts);
FRUIT: less than 1; DAIRY: 0

DIET NOTE

Now that I have told you how good this is, I have to also recommend that you save it for when you have reached your goal weight. It will use up calories, but a serving does not contain enough oats to constitute a starch serving, nor enough raisins for fruit, or oil and nuts for fat.

BAKING TIP

Dry and Wet Ingredients

When mixing a batter for quick breads and muffins, begin by mixing all the dry and all the wet ingredients separately, making sure each set of ingredients is thoroughly blended. This is the time to whisk or vigorously stir the dry ingredients so the flour will aerate, the leavening, salt, and spices will be evenly distributed, and any lumps in the sugar eliminated. Note that although brown sugar is technically a dry ingredient, I sometimes add it with the wet ingredients because it smoothes out more easily.

The wet ingredients should also be mixed briskly until thoroughly blended. Then, turn off your muscle and use a gentle, brief action to combine the dry and wet mixtures. Even if the batter appears lumpy, do not beat it smooth.

Overzealous mixing of dry and wet ingredients results in tough, rubbery muffins and breads that won't rise. Flours contain proteins called *gluten*, which give breadstuffs their structure. Gluten is only activated in the presence of liquid, and at that point it should be worked briefly for the most tender results.

Banana Bread

When you have a banana that has grown too ripe to be appetizing—that is, both the skin and fruit are black—seal it in a plastic bag (with the peel on) and freeze. When you have three ripe, frozen bananas, make banana bread. The frozen bananas will slip easily out of their skins.

3 very ripe bananas (6 to 7 inches each)
2 large eggs, beaten
½ cup extra virgin olive oil
1 teaspoon vanilla
2 cups whole wheat flour
¾ cup loosely packed brown sugar
1 teaspoon salt
1 teaspoon baking soda
½ teaspoon cinnamon
½ cup chopped walnuts

Equipment needed: 9 × 5-inch loaf pan, preferably nonstick

Preheat the oven to 350°F. Position a rack in the center of the oven.

Mash the bananas in a large bowl. The top of a sturdy wire whisk or a potato masher works well. Blend in the eggs. Stir in the olive oil and vanilla. Combine the flour, brown sugar, salt, baking soda, and cinnamon in a separate bowl. Mix well, being sure to break up any lumps of brown sugar. Stir in the walnuts.

Gently stir the dry ingredients into the wet ingredients just until combined. A plastic spatula works best. Do not overmix or the bread will not rise.

Pour the batter into the loaf pan and bake for 60 minutes, or until a cake tester comes out clean when inserted in the center of the loaf. Let the bread cool in the pan slightly, 5 to 10 minutes. Turn out onto a rack and cool thoroughly. Cut into 8 equal slices.

MAKES 8 SERVINGS
CALORIES: 380; STARCH: 1½ (flour); VEGETABLES: 0; FAT: 2 (1 each olive oil and nuts);
FRUIT: ¾ (banana); DAIRY: 0

Pumpkin Bread

*This moist bread is flavorful and good for you—the perfect answer to the healthy break-
fast question, especially for those who are used to eating high-calorie, negligible-nutrient
muffins in the morning. Although the bread has some vegetable (the pumpkin), some
fruit (raisins), and olive oil, it does not contain enough of any of them to include in the
food count, so count the starch and consider the calories.*

1½ cups whole wheat flour
1 cup loosely packed brown sugar
1 teaspoon baking soda
1 teaspoon cinnamon
½ teaspoon salt
1 cup canned pumpkin
½ cup extra virgin olive oil
2 large eggs, beaten
½ cup walnut pieces
½ cup raisins

Equipment needed: 9 × 5-inch loaf pan, preferably nonstick, or unglazed
ceramic stoneware. A stoneware pan will give the bread a particularly nice
crust.

Preheat the oven to 350°F. Position a rack in the center of the oven.

Combine the flour, brown sugar, baking soda, cinnamon, and salt in a
mixing bowl and stir with a fork or whisk until thoroughly combined. Be
sure to break up any lumps of brown sugar.

Combine the pumpkin, olive oil, and eggs in a separate bowl with ¼ cup
water. Beat until perfectly blended. Stir in the nuts and raisins.

Gently stir the dry ingredients into the liquid ingredients just until combined; a rubber spatula works best. Do not overmix or the bread will not
rise.

Pour the batter into the loaf pan and bake for 50 to 60 minutes, until a toothpick or cake tester inserted into the center comes out clean. Let the bread cool in the pan slightly, 5 to 10 minutes. Turn out onto a rack and cool thoroughly. Cut into 8 equal slices.

MAKES 8 SERVINGS

CALORIES: 370; STARCH: 1 (flour); VEGETABLES: less than 1; FAT: 2 (1 each olive oil and nuts); FRUIT: 0; DAIRY: 0

GOOD TO KNOW

Freezing Baked Goods

A dozen muffins or a loaf of fruit bread sitting on the kitchen counter are just begging for you to get into trouble. It is all too easy to start breaking off "just a small piece" until you have consumed half the batch or the entire loaf. So get them into the freezer for another day. After baking, let the pastries cool completely. Wrap each muffin or bread slice separately in plastic wrap or small plastic bags, then place in a larger plastic bag, squeeze out excess air, and freeze. Keep the plastic wrapping on when thawing to hold the moisture in. Bread slices will defrost quickly—in the time it takes you to exercise and shower; muffins take longer, so take one out of the freezer the night before. If you want to reheat a serving, place the unwrapped frozen pastry on a cookie sheet and heat in a 350°F oven or toaster oven until hot, 5 to 10 minutes.

Zucchini Bread

If you are a vegetable gardener, or your neighbor down the road is, summer probably means zucchini, and lots of it. This healthy version of an all-time favorite is a stellar way to use that bumper crop. Shred the zucchini in a food processor or with a hand grater.

2 cups shredded zucchini (2 small, 7 to 8 ounces each)
2 cups lightly packed brown sugar
1 cup extra virgin olive oil
3 large eggs, beaten
1 tablespoon vanilla extract
2 cups whole wheat flour
1 cup all-purpose flour
1 tablespoon cinnamon
2 teaspoons baking powder
1 teaspoon baking soda
1 teaspoon salt
1 cup coarsely chopped walnuts

Equipment needed: two 9 × 5-inch loaf pans, preferably nonstick

Preheat the oven to 350°F. Position a rack in the center of the oven.

Combine the zucchini, sugar, olive oil, eggs, and vanilla in a large bowl and stir until well blended. Combine the whole wheat flour, all-purpose flour, cinnamon, baking powder, baking soda, and salt in a separate bowl and stir vigorously until thoroughly mixed. Stir in the nuts.

Gently stir the dry ingredients into the zucchini mixture just until combined. A rubber spatula works best. Do not overmix or the bread will not rise.

Divide the batter between the two loaf pans and bake for 50 to 60 minutes, until a tester inserted in the center of the bread comes out clean. Let the breads cool in the pans slightly, 5 to 10 minutes. Turn the breads out onto a rack and cool thoroughly. Cut each loaf into 8 equal slices.

MAKES 16 SERVINGS
CALORIES: 330; STARCH: 1 (flour); VEGETABLES: ¼ (zucchini); FAT: 2 (1 each olive oil and nuts); FRUIT: 0; DAIRY: 0

Apricot-Almond Muffins

For some reason, I think of these as springtime muffins; perhaps it is the sunny color of the apricots. In truth, they can be made any time of year since they use dried fruit.

¾ cup dried apricots, finely chopped (about ¼-inch pieces)
1 cup whole wheat flour
1 cup all-purpose flour
1 cup loosely packed brown sugar
2 teaspoons baking powder
2 teaspoons cinnamon
½ teaspoon salt
½ cup slivered almonds
3 large eggs, beaten
¾ cup extra virgin olive oil
2 teaspoons almond extract

Equipment needed: 12-cup muffin tin

Preheat the oven to 375°F. Line the muffin tin with paper liners. Position a rack in the center of the oven.

Put the apricots in a small bowl. Pour 1 tablespoon hot water over them and let sit for 10 to 15 minutes, until plumped up.

Combine the whole wheat flour, all-purpose flour, sugar, baking powder, cinnamon, and salt in a bowl and mix together thoroughly. Make sure there are no lumps of brown sugar. Stir in the almonds. Combine the apricots, eggs, olive oil, and almond extract in a large bowl and mix well.

Make a well in the dry ingredients and pour in the liquid ingredients. Use a rubber spatula and a gentle touch to combine. The batter will not be smooth. Do not overmix or the muffins will not rise.

Divide the batter among the 12 muffin cups. Bake for 23 to 25 minutes, until the tops are golden brown and a cake tester or toothpick inserted into the top of a muffin comes out clean. Cool for about 5 minutes on a rack. Remove the muffins from the pan to finish cooling on the rack. The muffins will get soggy if they are allowed to cool completely in the pans.

VARIATIONS: Instead of apricots, can use the same amount of dried figs or dried dates. In both cases, use vanilla in place of the almond extract and try varying the nuts—pecans or walnuts, for example.

MAKES 12 SERVINGS
CALORIES: 315; STARCH: 1 (flours); VEGETABLES: 0; FAT: 1 ⅔ (1 olive oil, ⅔ nuts);
FRUIT: less than 1; DAIRY: less than 1

Blueberry Muffins

Use fresh blueberries or frozen ones that have been thawed. You can substitute slivered or sliced almonds for the whole nuts, but be careful not to process them beyond small pieces. Over-processing can turn the nuts into nut butter.

½ cup whole almonds
1 cup whole wheat flour
1 cup all-purpose flour
1 cup loosely packed brown sugar
2 teaspoons baking powder
2 teaspoons cinnamon
½ teaspoon salt
1 cup blueberries
3 large eggs, beaten
¾ cup extra virgin olive oil
1 teaspoon vanilla extract
1 teaspoon almond extract

Equipment needed: 12-cup muffin tin

Preheat the oven to 375°F. Line the muffin tin with paper liners. Position a rack in the center of the oven.

In a food processor fitted with the steel blade, process the almonds until they are small pieces; or chop by hand with a large (chef's) knife. Add the whole wheat flour, all-purpose flour, sugar, baking powder, cinnamon, and salt and process until well mixed, about 1 minute. Transfer the mixture to a bowl and stir in the blueberries.

Using a fork or whisk, mix the together the eggs, olive oil, and vanilla and almond extracts, beating well until thoroughly blended.

Make a well in the center of the dry ingredients and pour in the liquid ingredients. Use a rubber spatula to gently combine. The batter will not be smooth, but beat no further or the muffins will not rise.

Divide the batter among the 12 muffin cups. Bake for 23 to 25 minutes, until the tops are golden brown and a cake tester or toothpick inserted into the top of a muffin comes out clean. Cool for about 5 minutes on a rack. Remove the muffins from the pan to finish cooling on the rack. The muffins will get soggy if they are allowed to cool completely in the pans.

MAKES 12 SERVINGS
CALORIES: 300; STARCH: 1 (flours); VEGETABLES: 0; FAT: 1 ⅔ (1 olive oil, ⅔ nuts);
FRUIT: less than 1; DAIRY: 0

Morning Glory Muffins

These satisfying muffins are packed with so many good ingredients that Glory is a great name for them.

1 cup whole wheat flour
1 cup all-purpose flour
1 cup loosely packed brown sugar
2 teaspoons baking powder
2 teaspoons cinnamon
½ teaspoon salt
¾ cup coarsely chopped walnuts
½ cup shelled dry-roasted sunflower seeds
3 large eggs, beaten
1 cup crushed pineapple chunks with 2 tablespoons juice
¾ cup extra virgin olive oil
½ cup grated carrot (1 medium)
1 teaspoon vanilla extract
1 teaspoon almond extract

Equipment needed: 12-cup muffin tin

Preheat the oven to 375°F. Line the muffin tin with paper liners. Position a rack in the center of the oven.

Combine the whole wheat flour, all-purpose flour, sugar, baking powder, cinnamon, and salt in a large bowl. Mix together thoroughly, making sure there are no lumps of brown sugar. Stir in the walnuts and sunflower seeds to coat with the mixture. Mix the together the eggs, pineapple, olive oil, carrot, and vanilla and almond extracts in a separate bowl.

Make a well in the center of the dry ingredients and pour in the liquid ingredients. Use a rubber spatula to gently combine. The batter will not be smooth. Do not overmix or the muffins will not rise.

Divide the batter among the 12 muffin cups. Bake for 27 to 29 minutes, until the tops are golden brown and a cake tester or toothpick inserted into the top of a muffin comes out clean. Cool for about 5 minutes on a rack. Remove the muffins from the pan to finish cooling on the rack. The muffins will get soggy if they are allowed to cool completely in the pans.

MAKES 12 SERVINGS
CALORIES: 345; STARCH: 1; VEGETABLES: less than 1; FAT: 2 ⅔ (1 each olive oil and nuts, ⅔ sunflower seeds); FRUIT: less than 1; DAIRY: less than 1

Dried Cranberry and Almond Muffins

Think Thanksgiving morning! What could be more welcome than a muffin with the traditional fruit of the day? If you want to think Christmas instead, substitute pistachios for the almonds.

1 cup whole wheat flour
1 cup all-purpose flour
1 cup loosely packed brown sugar
2 teaspoons baking powder
1 teaspoon cinnamon
½ teaspoon ground cloves
½ teaspoon salt

½ cup slivered almonds
1 cup dried cranberries
3 large eggs, beaten
¾ cup extra virgin olive oil
1 teaspoon vanilla extract
1 teaspoon almond extract

Equipment needed: 12-cup muffin tin

Preheat the oven to 375°F. Line the muffin tin with paper liners. Position a rack in the center of the oven.

Combine the whole wheat flour, all-purpose flour, sugar, baking powder, cinnamon, cloves, and salt in a bowl and mix together thoroughly. Make sure there are no lumps of brown sugar. Stir in the almonds and cranberries. Beat together the eggs, olive oil, and vanilla and almond extracts in a separate bowl until thoroughly blended.

Make a well in the center of the dry ingredients and pour in the liquid ingredients. Use a rubber spatula to gently combine. The batter will not be smooth and it will be thick. Do not overmix or the muffins will not rise.

Divide the batter among the 12 muffin cups. Bake for 23 to 25 minutes, until the tops are golden brown and a cake tester or toothpick inserted into the top of a muffin comes out clean.

Cool for about 5 minutes on a rack. Remove the muffins from the pan to finish cooling on the rack. The muffins will get soggy if they are allowed to cool completely in the pans.

MAKES 12 SERVINGS
CALORIES: 320; STARCH: 1; VEGETABLES: 0; FAT: 1⅔ (1 olive oil, ⅔ nuts); FRUIT: less than 1; DAIRY: less than 1

Peanut Butter–Oatmeal Bars

Why spend money on store-bought cereal bars that usually aren't as healthy as they claim? One of these bars, along with a serving of fruit, provides all you need for a perfectly healthy PBOO breakfast. And they are a snap to make. One bowl is all you need—even kids can do it. Let the baked bars cool completely before eating and store the pieces, individually wrapped, in the freezer.

⅔ cup whole wheat flour
1 cup loosely packed brown sugar
1 cup dry old-fashioned (rolled) oats
1 teaspoon salt
¼ teaspoon baking soda
½ cup extra virgin olive oil
½ cup natural peanut butter
1 large egg, beaten
1 tablespoon 1% milk

Equipment needed: 9 × 9-inch baking pan, preferably nonstick

Preheat the oven to 375°F.

Combine the flour, sugar, oats, salt, and baking soda in a large bowl. Blend together the olive oil, peanut butter, egg, and milk in another bowl.

Make a well in the center of the dry ingredients and add the liquid ingredients. Using a rubber spatula, combine the ingredients together.

Press the mixture into the baking pan and bake for 14 to 15 minutes, until the edges are browned. Cool in the pan. Cut into 8 bars.

MAKES 8 SERVINGS
CALORIES: 360; STARCH: 1 (flour and oats combined); VEGETABLES: 0; FAT: 2 (1 each olive oil and peanut butter); FRUIT: 0; DAIRY: less than 1

COOKING TIP

Old-Fashioned Oats

The recipes in this book that use oats were all tested with 5-minute, old-fashioned oats, also known as rolled oats. These are different than quick-cooking oats and instant oatmeal, which are precooked and dried so they only need to meet the hot water to be ready to eat, and also often have added flavorings, sugar, and salt. Quick-cooking oats and instant oatmeal are not good substitutes for rolled oats unless you are particularly fond of gooey.

Steel-cut oatmeal is nice to have on hand, although it will take much longer to cook than the 5-minute variety. Cook up a big pot of steel-cut oatmeal on the weekend, and reheat it during the week for breakfast or to use in any of the recipes calling for oatmeal. Buy it in bulk at a natural food store to avoid paying for the fancy tin at the grocery store.

Breakfast Tortilla

When you want something special for a late breakfast or Sunday brunch, this simple version of Huevos Rancheros is a great choice—for family and friends as well as yourself. It is also a perfect lunch. If you want your beans to have the texture of refried beans, mash them after letting them heat with the salsa. A potato masher works great, but the back of a fork will do the trick.

1 tablespoon extra virgin olive oil, plus more for heating the tortilla
  and cooking the egg
¼ cup chopped onion
Salt
½ cup cooked black beans (page 128) or canned beans, drained and
  rinsed
¼ cup favorite salsa
6-inch whole wheat tortilla
1 large egg
2 tablespoons shredded Cheddar or Monterey Jack cheese

Equipment needed: two 8-inch skillets (one to cook the bean mixture, the other to heat the tortilla and then cook the egg)

Heat the olive oil in a medium skillet over medium heat. Stir in the onion, season with salt, and cook until soft, 3 to 5 minutes. Stir in the beans and cook over medium heat a few minutes to warm them. Add the salsa, give a good stir to combine, and cook until the mixture is heated through. Keep over low to medium heat while heating the tortilla and cooking the egg.

Lightly oil the second skillet and place over medium-high heat. Lay the tortilla in the pan and cook, turning once, until warmed through. Remove from the pan and keep warm.

Add a little more oil to the tortilla pan and place over medium heat. Add the egg and cook until set, sunny side up. Sprinkle with the cheese, cover, and cook briefly, until the cheese melts. Meanwhile, put the warm tortilla on a plate and top with bean mixture. Carefully slide the egg out of the pan and onto the beans. Serve at once.

MAKES I SERVING
CALORIES: 455; STARCH: 2 (I beans, I tortilla); VEGETABLES: I (½ onion, ½ salsa); FAT: I (olive oil); FRUIT: 0; DAIRY: ½ (cheese)

INCREASING THE RECIPE
To make this recipe for more than one person, increase the bean ingredients by the number of people you want to feed and cook them all at once. Cook the tortillas individually (you may need to add more oil as you cook them); wrap in foil to keep warm. Fry as many eggs as you can handle at a time.

Breakfast Egg and Vegetable Sandwich

Frozen, pre-sliced peppers are a godsend to the hurried breakfast cook. No need to de-frost them; just drop into the heated oil and they will do it all by themselves. Frozen pep-pers are available in both bags and boxes, usually packaged with red, green, and yellow varieties together. I like the bagged type. Although the boxed frozen peppers will break apart like frozen peas, it is easier to grab a handful from the bag.

1 tablespoon extra virgin olive oil
½ cup sliced frozen or fresh bell peppers, red, green, and/or yellow
    (about ½ medium fresh pepper)
½ cup thinly sliced mushrooms
Salt and pepper
1 large egg
2 slices whole wheat bread

Heat the olive oil in a small sauté pan over medium-low heat. Add the peppers and mushrooms and season with salt and pepper. Cook until the mushrooms release and then reabsorb their juices and are lightly browned, 5 to 10 minutes.

Break the egg on top of the vegetables and cook until the white is opaque. Carefully flip the egg and vegetables and cook for 2 to 3 minutes longer for a runny yolk, or 4 to 5 minutes longer for a firmer egg. Mean-while, toast the bread. Slip the egg and vegetables onto the toast and serve.

VARIATIONS: There are a number of ways you can vary this sandwich: Use 1 cup of any vegetable—chopped broccoli, peppers, or mushrooms are great—or combination of vegetables. Cut the vegetables small enough so there is space between the pieces for the heat of the pan to reach the egg. You can also sprinkle ¼ cup (1 ounce) of grated cheese on top once you have flipped over the egg and vegetables. In that case, add 1 serving of dairy and 110 calories to your daily count.

MAKES I SERVING
CALORIES: 370; STARCH: 2 (toast); VEGETABLES: 2 (I each peppers and mushrooms); FAT: I (olive oil); FRUIT: 0; DAIRY: 0

Soups, Sandwiches, and Salads

Although the recipes in this section make fine dinners, they are particularly well-suited for lunch. If you regularly eat lunch away from home, preparing your own food is important. The more meals you purchase out, the harder it will be to control what you eat. Bringing your lunch with you does mean that you have to plan ahead—shopping for foods that make portable lunches and containers for transporting them. A plastic container is good for sandwiches and salads. A thermos or microwaveable plastic bowl is good for soups.

Since mornings can be rushed, get in the habit of preparing lunch the night before, perhaps when you are making dinner. Depending on the bread, vegetable sandwiches could become soggy waiting overnight in the refrigerator or even during the trip from home to work, so I recommend wrapping the vegetables and the bread separately and putting them together when you are ready to eat.

The soups can all be made two or three days ahead of time. They almost always improve with age. For longer keeping, divide the soup into serving sizes and freeze in individual containers.

Mixing It Up

Want soup *and* a sandwich or salad for lunch or supper? This is not a problem. A serving size of most of the soups is a generous 2 to 2½ cups. If you cut that serving in half, you can still enjoy a satisfying cup (or slightly more) of soup (1 fat) with half a serving of a 2-fat-count sandwich or salad, making the total fat 2 servings. If you select a sandwich or salad that has just 1 serving of fat, you could have the entire serving with a ½ cup of the soup.

GOOD TO KNOW

About Broth

Most of our recipes call for vegetable broth, a few for a combination of vegetable broth and water. You can, of course, make your own vegetable broth. You will find fat-free recipes in many vegetarian cookbooks and on the Internet. Water alone is always an option but the soup will not have as much flavor as it will with the vegetable broth. Commercial vegetable broth is sold in markets in cans and boxes (ready to use) and in dehydrated (bouillon) cubes and powders, which should be reconstituted by adding 1 cup water to 1 cube or to 1 heaping teaspoon powder. Be sure to stir them well so they are thoroughly dissolved.

Low-sodium broth is the best choice, because soups usually reduce somewhat when they are cooking, and the salt could become too concentrated. You will have better control over the saltiness if you start with low-sodium broth. "Needs a little less salt" can't help a soup you are about to eat!

There is another way you can balance your meal to have both soup and sandwich or salad. Almost all the soups begin with a ½ cup of olive oil—or 2 fat allowances per serving. I wrote them that way for the simplicity of the diet study so a serving would fit the ideal allowances for lunch. They work that way and taste good as written. But all the soups work just as well when started with a ¼ cup of olive oil—or 1 fat per serving. If you like the ease of knowing that the soup you are eating is all you need for the ideal lunch, then use the ½ cup oil; otherwise use a ¼ cup and have the side you want or add another serving of fat into the meal elsewhere—perhaps nuts on a salad, half an English muffin spread with peanut butter, or a toasted slice of whole wheat Italian bread brushed with olive oil.

Choosing Bread Products

Have you taken a good look lately at the bread aisle in your grocery store? Actually, most large markets no longer have just an aisle. They have sections—a section for cellophane-wrapped loaves and rolls, a section for unpackaged "freshly baked" breads such as French and Italian, a section for ethnic-inspired chapati, tortillas, *lavash,* and pita bread, a frozen section for bagels. This is all good. Really good. The more you vary the breads you eat, the more interesting your diet will be. But how to choose?

Right off the bat, you have to determine if the product is as good for you as it could be. Don't look at the claims on the front of the package; along with the proliferation of breads came the explosion of package claims of how healthy the breads are. Maybe, maybe not. What we know for sure is that fiber can do lots for you, so turn the package over and check the nutrition label. The first ingredient should be whole wheat, whole grain, whole rye, or any grain that is "whole." That's where the fiber and the phytonutrients are. Naturally, 100 percent "whole" would be the best choice. Products usually list the 100 percent on the label, but you can figure it out by looking at the other ingredients: if the first ingredient is a whole grain and there are no other grains listed, then most likely it is a 100 percent whole grain product.

After that, it gets a bit tricky. In order for a product to be listed as "whole grain," the USDA requires that 51 percent, just over half, of the ingredients by weight be whole grain. But let's say the first ingredient listed is whole wheat flour and it's followed by two or three other types of flour that are not whole grain. The total weight of the other flours could exceed the weight of the whole wheat, in which case it's not a whole grain bread and should not be labeled as such—but it happens. Shame! If you are comparing two products and not sure which is better, look for a higher count of dietary fiber. The next thing to check is the serving size and the number of calories in that serving. This information is listed at the top of the nutrition label. You are allowed 6 to 7 starches per day, each one being between 80 and 100 calories. If the label states that 1 serving has about 100 calories, then a serving of the product constitutes just one of the 7 starches allowed for the day. If, on the other hand, the calorie count is 160 for 1 serving, that would mean 2 allowances. That does not mean you always have to choose the lower calorie product. Just be sure and count it properly in your allowances. Look carefully at what the package lists as a serving size or you could wind up eating more calories than you realize. In most cases, if a package has eight pieces of a breadstuff, one serving is one of those eight pieces. But for some products, such as *lavash* (*lahvosh*) and large pita breads or large tortillas, the serving size is just half a piece and the calories listed are for a half, not a whole, piece.

And what do you do when there is no package with labeling to check? You have to depend on weight. In general, breads and rolls without the addition of nuts or raisins or seeds or other such enhancements have about 75 calories per ounce (although English muffins have 60 or less). A typical sandwich roll weighs 3 to 5 or more ounces. At 75 calories an ounce, one roll can "cost" you from 225 calories (a little over 3 starch servings) to a whopping 375 calories—more servings than you should squander on one meal's worth of starch. With long loaves of Italian or French breads, weigh the entire loaf and mentally divide it into slices to determine how big a serving would be and if you want it.

When we have given you the number of calories and the starch allowance following a recipe, we have based it on the bread we used. Your piece may have more or less calories than the one we used, so your recipe will use up a greater or lesser number of your allowances. Determine the count by the number of calories per piece in what you use—80 to 100 is 1 starch, 160 to 200 is 2 starches, and so forth.

We love finding bread products which constitute just 1 or at most 2 starch servings, and if you spend a little time turning packages over and reading the labels, you'll discover a host of choices. There are some we especially recommend because they are calorie efficient, work well for sandwiches "on the go," and taste good. Sandwich Thins, put out by Arnold Bread, have only 100 calories (1 serving) and are just the right size for a sandwich. In the recipes they are referred to as *sandwich rounds*. They look a bit like a crumpet or an English muffin, with the top and bottom halves split in two to accommodate a filling in the middle. Wraps—*lavash* (*lahvosh*), chapati, tortillas—come in a variety of sizes and flavors such as spinach, tomato, and chipotle. Depending on the brand, wraps are usually low enough in calories to count as 1 or at most 2 starches and they hold up well in sandwiches made with moist fillings. Most small, 4-inch pita breads (pockets) are 1 serving; 6½-inch pitas are 2 servings. Although the small ones may not accommodate an entire sandwich filling, they are delicious when they become Pita Chips (page 104) to serve with soup or dips.

Crackers are not a bread but close enough to be included here. Like breads they should be whole grain—once you have determined that they

GOOD TO KNOW

Finishing Soups

Here are a few simple finishes for soups that will give them a little extra something.

1. Sprinkle 2 to 4 tablespoons grated Parmesan, Romano, or Asiago cheese on top of each serving of hot soup. Add ½ to 1 dairy to food count.
2. Make the soup with ¼ cup olive oil instead of ½ cup and drizzle a tablespoon of your best extra virgin olive oil on top of each serving. The heat of the soup will accentuate the flavor of the oil. Italians call this "anointing" or "christening" the soup and always drizzle it on in the shape of the letter "C." Fat count for each serving remains at 2.
3. Stir 2 tablespoons of low-fat or nonfat plain, unsweetened yogurt into a serving of hot soup. Especially good in pureed soups since it adds creaminess. Count it as ¼ dairy.
4. Make the soup with ¼ cup of olive oil instead of ½ cup and finish each serving with 2 tablespoons of Arugula Pesto (page 154) or Basil Pesto (page 153). This is especially good in vegetable soups. Fat count is a total of 1½ servings.
5. Sprinkle chopped fresh herbs on any of the soups. Basil, parsley, and chives particularly have flavors that will be instantly released when they meet the hot soup. No additional food count.

are, look to see if there is fat listed. (If it says trans fats or partially hydrogenated fat, drop the product on the floor and if no one is looking stomp on it.) What you want to see is "extra virgin olive oil" or "olive oil," which, even though it cannot compete with extra virgin, is still better than any vegetable oil. Once in a while, occasionally, not very often, rarely when you find yourself having to purchase and eat whole wheat crackers that are made with vegetable oil, only choose those that have no more than 5 grams of fat per ounce.

Curried Butternut Squash and Apple Soup

A blender gives this soup a rich, creamy texture. You can use a food processor, but the soup will be less smooth. Either way, be sure to cool the soup slightly before pureeing or you may end up painting your ceiling a lovely orange-yellow color.

¼ cup extra virgin olive oil
1 cup chopped onion (1 medium)
½ cup chopped peeled carrot (1 medium)
½ cup chopped celery (1 medium stalk)
2 medium garlic cloves, minced
Salt and black pepper
1 tablespoon curry powder
4 cups butternut squash cut into ½-inch cubes (2¼-pound squash, peeled and seeded)
2 cups chopped tart apples, such as Granny Smith (2 small, peeled and cored)
4 cups low-sodium vegetable broth
1 cup apple cider
½ cup unflavored, unsweetened low-fat or nonfat yogurt

Equipment needed: blender (immersion or free standing)

Heat the olive oil in a large soup pot over medium heat. Add the onion, carrot, and celery, season lightly with salt, and cook until the onion is translucent but not brown, about 10 minutes. Stir in the curry powder and continue to cook for 5 minutes to release its flavor. Add the squash and apples, season with salt and pepper, and stir to coat with the oil and mix with the vegetables. Cook, stirring occasionally, for 15 minutes.

Pour in the vegetable broth and ½ cup of the cider. Bring the soup to a boil, then reduce the heat to medium. Cover and simmer until the squash and apples are tender, about 30 minutes. Cool slightly.

Meanwhile, pour the remaining ½ cup cider into a small saucepan and boil until it has reduced to ¼ cup, about 5 minutes. Cool and then whisk it into the yogurt. The cider yogurt may be made a day ahead and kept covered in the refrigerator.

Puree the soup with an immersion blender or by working in batches in a blender. When ready to serve, reheat the soup and serve each bowl with a dollop of the cider yogurt.

MAKES 4 SERVINGS, ABOUT 2 CUPS EACH
CALORIES: 380; STARCH: 0; VEGETABLES: 3 (¼ carrot, ¼ celery, ½ onion, 2 butternut squash); FAT: 1 (olive oil); FRUIT: 1¼ (apple, apple juice); DAIRY: ¼ (yogurt)

Super Cruciferous Soup

Broccoli and cauliflower are cruciferous vegetables, which have impressive, super-cancer-fighting phytonutrients called glucosinolates (see page 242). This soup is a fine way to get two varieties of cruciferous vegetables in one sitting. If you use all frozen vegetables, the recipe will cook up faster than it will with fresh.

½ cup extra virgin olive oil
2 cups chopped broccoli, fresh or thawed frozen
2 cups chopped cauliflower, fresh or thawed frozen
Salt and black pepper
4 cups baby spinach, or ⅔ cup chopped thawed frozen spinach
6 cups low-sodium vegetable broth
1½ pounds unpeeled baking potatoes (4 small), scrubbed clean and
    cubed (4 cups)

Heat the olive oil over medium heat in a soup pot. Stir in the broccoli and cauliflower and season with salt and pepper. Reduce the heat to medium-low and cook until the vegetables begin to soften, 15 to 20 minutes. Add the spinach, season lightly with salt, and cook until the spinach has wilted; frozen spinach should be cooked for about 10 minutes to absorb the olive oil.

Pour in the vegetable broth, raise the heat, and bring the broth to a boil. Add the potatoes, cover, and reduce the heat to medium. Cook at a slow boil until the potatoes can be easily pierced with a fork, about 10 minutes.

VARIATIONS: Two or three whole, fat cloves of garlic, halved or minced, cooked with the broccoli and cauliflower are a fine addition. If you like a little "fire" in your soup, use ¼ teaspoon hot red pepper flakes in place of the black pepper.

MAKES 4 SERVINGS, ABOUT 2½ CUPS EACH
CALORIES: 425; STARCH: 2 (potatoes); VEGETABLES: 3 (1 each broccoli, cauliflower, and spinach); FAT: 2 (olive oil); FRUIT: 0; DAIRY: 0

Simple Vegetable Soup

I like this soup, and not just because it is simple. It also has a good variety of vegetables, hence a good variety of phytonutrients, which are pretty much all absorbed by our bodies because of the olive oil.

½ cup extra virgin olive oil
1 cup chopped red onion (1 small)
Salt and black pepper
1 cup chopped broccoli, fresh or thawed frozen
1 cup chopped carrots (2 medium)
4 cups baby spinach, or ⅔ cup thawed frozen chopped spinach
1 (28-ounce) can crushed tomatoes
6 cups water or low-sodium vegetable broth
1½ pounds unpeeled baking potatoes (4 small), scrubbed clean and
    cubed (4 cups)

Heat the olive oil in a soup pot over medium-low heat. Add the onion, season with salt and pepper, and cook for 10 minutes. Stir in the broccoli, carrots, and spinach and season with salt and pepper. Cook until the vegetables are tender, 25 to 30 minutes.

Add the tomatoes and vegetable broth. Cover and bring the soup to a boil over medium-high heat. Add the potatoes and reduce the heat slightly. Partially cover and simmer until the potatoes are tender, 10 to 15 minutes.

MAKES 4 SERVINGS, ABOUT 3 CUPS EACH

CALORIES: 450; STARCH: 2 (potatoes); VEGETABLES: 3½ (½ each onion, broccoli, and carrots; 1 each spinach and tomato); FAT: 2 (olive oil); FRUIT: 0; DAIRY: 0

Ribollita

This is an easy version of a classic Tuscan soup. Ribollita means re-boiled or twice cooked, referring to the method Italians use to make it. They add stale bread to leftover minestrone, reheat it, and ecco!—a whole new soup. In any language, it's a good way to use stale bread. Ribollita is at its best if you let the soup mellow a few hours or overnight without the bread, and then finish cooking with the bread just before serving. The bread should be one or two days old—dried out but not rock hard. Leave it out of its wrapping overnight so it will dry out. If you want to freeze the soup, do it before adding either the bread or the beans.

½ cup extra virgin olive oil
1 cup chopped red onion (½ medium)
Salt and black pepper
4 cups shredded Savoy cabbage (about ½ pound)
1 cup chopped celery (2 medium stalks, include leaves if possible)
1 cup thinly sliced carrots (2 medium)
1 (28-ounce) can crushed tomatoes
4 cups low-sodium vegetable broth
2 cups cooked cannellini beans (page 128), or canned beans, drained
    and rinsed
3 to 4 (2-inch) sprigs fresh thyme
4 ounces day-old whole wheat bread, torn or cut into 1-inch pieces

Heat the olive oil in a soup pot over medium heat. Stir in the onion, season lightly with salt, and cook until translucent, about 10 minutes. Stir in the cabbage and cook for 5 minutes. Add the celery and carrots and cook for 5 minutes. Add the tomatoes, broth, and beans, then drop in the thyme. Turn up the heat and cover. When the soup comes to a boil, reduce the heat to medium-low. Partially cover and simmer for 15 to 20 minutes, until the vegetables are tender; longer cooking will only improve the flavors. Let the soup cool, then cover and refrigerate overnight or let rest at room temperature for several hours.

When ready to serve, bring the soup to a boil and stir in the bread. Return to a boil, reduce the heat, and simmer for about 8 minutes to blend the flavors.

VARIATIONS: Cook 3 or 4 minced garlic cloves with the onions. Sprinkle the finished soup with grated Parmesan cheese—measure and count the dairy. Start the soup with just ¼ cup olive oil and drizzle 1 tablespoon of your best extra virgin olive oil on each serving. Instead of adding torn bread to the soup, toast a piece of Italian whole wheat bread, brush with olive oil, and rub the oiled side with a cut clove of garlic; then place the toast in the bottom of a soup bowl, pour the hot soup on top, and sprinkle with chopped parsley.

MAKES 4 SERVINGS, ABOUT 2½ CUPS EACH
CALORIES: 565; STARCH: 2 (1 each beans and bread); VEGETABLES: 4½ (½ each onion, carrots, and celery; 1 cabbage, 2 tomatoes); FAT: 2 (olive oil); FRUIT: 0; DAIRY: 0

Black Bean, Greens, and Barley Soup

This hearty soup is perfect on a cold day.

½ cup extra virgin olive oil
1 cup red onion, chopped (½ medium)
Salt and black pepper
1 ⅓ cups thawed frozen chopped kale or turnip greens
6 cups low-sodium vegetable broth
2 cups cooked black beans (see page 128), or canned beans, drained and
    rinsed
2 cups cooked barley (see page 138)
½ cup loosely packed grated Parmesan, Romano, or Asiago cheese
    (optional)

Heat the olive oil in a large soup pot over medium heat. Stir in the onion, season lightly with salt and pepper, and cook until translucent but not brown, about 10 minutes. Toss in the greens, season lightly, and reduce the heat to low. Cook gently until the greens are tender, about 15 minutes.

Add the vegetable broth, beans, and barley. Cook until heated through, about 10 minutes. Sprinkle 2 tablespoons of Parmesan cheese, if using, on top of each serving.

VARIATIONS: I use kale and turnip greens but any dark leafy vegetable would be equally as good. Try collards or mustard greens. Three cloves of chopped garlic cooked with the onions is also a good addition. The frozen vegetables make a quick soup, but you can substitute fresh. In that case you will need about 2 ⅔ cups of coarsely chopped greens.

MAKES 4 SERVINGS, ABOUT 2½ CUPS EACH
CALORIES: 510; STARCH: 2 (1 each beans and barley); VEGETABLES: 2½ (½ onion, 2 greens); FAT: 2 (olive oil); FRUIT: 0; DAIRY: ½ (optional cheese)

Tomato and Kidney Bean Soup

This is an easy, satisfying soup that freezes well, so think about increasing the recipe and freezing it in 2-cup serving sizes.

½ cup extra virgin olive oil
½ cup red onion slices (½ small)
½ cup diced celery (1 medium)
Salt and black pepper
2 cups baby spinach or ⅔ cup thawed frozen spinach
1 (28-ounce) can crushed tomatoes
4 cups low-sodium vegetable broth
2 cups cooked kidney beans (see page 128), or 2 cups canned beans,
    drained and rinsed
1 small sprig fresh rosemary

Heat the olive oil in a soup pot over medium heat. Add the onion and celery, season lightly with salt, and cook until the vegetables begin to soften, 8 to 10 minutes. Drop in the spinach, reduce the heat slightly, and cook just until the spinach has wilted; frozen spinach should be cooked for 10 minutes so it absorbs the olive oil.

Add the tomatoes, broth, beans, and rosemary and season lightly with salt and pepper. Simmer for 15 or 20 minutes to blend the flavors.

VARIATIONS: Keeping to the proportions and following the technique below, there are a number of simple and savory ways to vary this recipe. Substitute kale, Swiss chard, or shredded red cabbage for the spinach. Cook 2 to 3 chopped garlic cloves with the onions. Use different beans. Replace the rosemary with thyme or a bay leaf. Finish the soup with Arugula Pesto (page 154) or Basil Pesto (page 153).

MAKES 4 SERVINGS, ABOUT 2 CUPS EACH
CALORIES: 440; STARCH: 1 (beans); VEGETABLES: 3 (¼ each onion and celery, ½ spinach, 2 tomatoes); FAT: 2 (olive oil); FRUIT: 0; DAIRY: 0

Roasted Vegetable and Lentil Soup

The vegetables for this stew are first roasted at high heat, which gives them a compelling, concentrated flavor. For the tomatoes, I use those labeled "Italian-style" because they contain seasonings that work well with the soup ingredients—garlic, oregano, and basil. If you can't find Italian-style tomatoes, add your own seasonings: Roast 3 or 4 whole cloves of garlic and 1 teaspoon dried or 1 tablespoon fresh oregano with the eggplant and onions, and roast 2 tablespoons chopped fresh or 2 teaspoons dried basil with the tomatoes.

1 large eggplant, about 1 pound
1 medium red onion, peeled and quartered (2 cups)
1 cup green or red bell pepper chunks (1 medium)
Salt
¼ cup plus 2 tablespoons extra virgin olive oil
1 (14 ½-ounce) can Italian-style tomatoes
1 cup lentils
4 to 6 cups low-sodium vegetable broth or water, or a combination

Preheat the oven to 450°F.

Cut off the top and the bottom of the eggplant. Do not peel. Cut into large chunks and place in a roasting pan. Add the onion and pepper and season with salt. Pour ¼ cup of the olive oil over them. Roast for 15 minutes. Use a fork to turn the vegetables over. Roast for 15 minutes longer, or until browned.

Meanwhile, stir the remaining 2 tablespoons olive oil into the tomatoes and place in a small baking dish. Season lightly with salt and pepper. Roast the tomatoes alongside the vegetables for about 20 minutes, or until most of the liquid has evaporated and the tomatoes start to brown. Periodically check the tomatoes, as they can burn easily. Remove both pans from the oven and let the vegetables cool slightly.

Peel the skin from the eggplant and put the flesh and the onion (but not the peppers) in a food processer fitted with a steel blade. Process until smooth. Transfer to a soup pot along with the lentils and broth. Bring to a boil and reduce the heat. Cover and simmer for 30 to 35 minutes, until the lentils are soft. Stir in the roasted tomatoes and the peppers and serve.

VARIATIONS: If you use the lesser amount of broth, you will have more of a stew than a soup—and a good one at that. For a different version, use all the broth and puree the tomatoes and peppers with the eggplant and onions. Either way, the soup can be finished with a tablespoon or two of plain, unsweetened yogurt.

MAKES 4 SERVINGS, ABOUT 1½ CUPS EACH
CALORIES: 440; STARCH: 2 (lentils); VEGETABLES: 4 (½ peppers, 1 each onion and tomato, 1½ eggplant); FAT: 2 (olive oil); FRUIT: 0; DAIRY: 0

Lentil Soup

Make this soup with either the grayish-brown variety of lentils or with red lentils. The red ones will make a thicker soup, but they are not always easy to find. If your regular grocery store doesn't have them, look in Middle Eastern or Indian markets.

½ cup extra virgin olive oil
2 cups chopped red onion (1 medium)
Salt and black pepper
1 cup chopped carrots (2 medium)
1 cup chopped celery (2 medium stalks, leaves included if possible)
6 cups low-sodium vegetable broth
1 cup lentils
2 to 3 sprigs fresh thyme
2 bay leaves
2 cups cooked brown rice (see page 138)

Heat the olive oil in a large soup pot over medium heat. Stir in the onion, season lightly with salt, and cook until softened, about 5 minutes. Add the carrots and celery, season lightly, and reduce the heat to low. Simmer for 25 minutes.

Add the broth, lentils, thyme, and bay leaves. Bring to a boil and reduce the heat to low. Cover and cook until the lentils are soft, about 45 minutes. Stir in the brown rice and cook just until heated through. Discard the thyme sprigs and bay leaves before serving.

MAKES 4 SERVINGS, ABOUT 2½ CUPS EACH
CALORIES: 610; STARCH: 2 (1 each lentils and rice); VEGETABLES: 2 (½ each onion and celery, 1 onion); FAT: 2 (olive oil); FRUIT: 0; DAIRY: 0

Lentil Stew with Spinach and Red Potatoes

This is my version of a Mediterranean–Middle Eastern stew that I enjoyed in Athens. In between rapturously savored spoonfuls, I jotted down notes on the flavors, then later recreated the soup at home with the help of a Greek cookbook.

½ cup extra virgin olive oil
2 garlic cloves, minced
¼ teaspoon cayenne pepper
Salt
4 cups low-sodium vegetable broth
1 cup brown lentils
12 ounces unpeeled red-skinned potatoes (about 4), scrubbed clean
    and cut into ½-inch pieces
8 cups baby spinach
1 teaspoon lemon zest
1 tablespoon lemon juice
4 tablespoons feta cheese

Heat the olive oil in a soup pot over medium-low heat. Stir in the garlic and cook until golden, but not brown. Season with cayenne pepper and salt and cook for another minute to release the flavor of the cayenne. Add the vegetable broth and lentils. Bring to a boil and reduce the heat. Cover and simmer for 10 minutes.

Add the potatoes and cook, uncovered and stirring occasionally, until the lentils and potatoes are tender, about 15 minutes. Stir in the spinach, lemon zest, and lemon juice. Cover and simmer until the spinach has wilted, about 5 minutes. Sprinkle each serving with 1 tablespoon feta.

MAKES 4 SERVINGS, ABOUT 2 CUPS EACH
CALORIES: 530; STARCH: 3 (1 potatoes, 2 lentils); VEGETABLES: 2 (spinach); FAT: 2 (olive oil); FRUIT: 0; DAIRY: ¼ (feta)

End-of-the-Summer Vegetable Stew

A very tasty way for gardeners to use the last of the tomatoes and the first of the egg-plant. Starting the eggplant first in the microwave speeds up the cooking in the pot; and, since eggplant is piggish about oil, the initial cooking slows down its appetite so there is enough oil to cook the other vegetables.

4 cups diced unpeeled eggplant (about 1 pound)
½ cup extra virgin olive oil
2 cups fresh green beans cut into 2-inch pieces (about 8 ounces)
2 cups diced red and/or green bell peppers (2 medium)
2 cups diced red onion (1 medium)
Salt
2 cups diced fresh tomatoes (about 9 ounces) or 2 cups canned tomatoes
    with juices
2 teaspoons paprika
2 cups low-sodium vegetable broth
1½ pounds unpeeled baking potatoes (4 small), scrubbed clean and
    cubed (4 cups)

Cook the eggplant in the microwave on high for 3 minutes.

Heat the olive oil in a large soup pot over medium heat. Stir in the egg-plant, green beans, bell peppers, and onion. Season lightly with salt and cook, stirring occasionally, for 20 minutes. Add the tomatoes and paprika and cook until heated through, about 10 minutes longer. Add the broth and potatoes and bring to a boil. Reduce the heat, cover, and simmer until the potatoes are tender, 15 or 20 minutes.

MAKES 4 SERVINGS, 2½ CUPS EACH
CALORIES: 475; STARCH: 2 (potatoes); VEGETABLES: 6 (1 each green beans, pepper, onion, and tomato; 2 eggplant); FAT: 2 (olive oil); FRUIT: 0; DAIRY: 0

Vegetarian Chili

Think a really good chili has to have meat? Well, it doesn't. This family-pleasing vegetarian version is rich and satisfying without it. If you like, you can spoon your serving of chili over cooked brown rice. Depending on how many starches you have planned for the day, use 1 or 1½ cups rice (2 or 3 servings).

½ cup extra virgin olive oil
2 cups chopped red onion (1 medium)
1 tablespoon chili powder
2 teaspoons ground cumin
2 teaspoons dried oregano or 1 scant tablespoon chopped fresh oregano
Salt
2 cups drained canned corn or thawed frozen corn
1 (28-ounce) can crushed tomatoes
1¾ cups cooked kidney beans (see page 128), or canned beans, drained
   and rinsed
1¾ cups cooked black beans (see page 128), or canned beans, drained
   and rinsed
1¾ cups cooked pinto beans (see page 128), or canned beans, drained
   and rinsed

Heat the olive oil in a large soup pot over medium heat. Stir in the onion and season with the chili powder, cumin, oregano, and salt. Cook, stirring occasionally, until the onion is translucent, about 10 minutes. Add the corn, stir it around, and cook for 5 minutes so it picks up the flavors of the onion and herbs. Add the tomatoes and all the beans and stir to combine. Reduce the heat to medium-low and simmer 45 minutes to deepen and blend the flavors.

MAKES 8 SERVINGS, 1 CUP EACH
CALORIES: 355; STARCH: 1⅓ (beans); VEGETABLES: 1½ (½ corn, 1 tomatoes);
FAT: 1 (olive oil); FRUIT: 0; DAIRY: 0

Cheese, Pepper, and Onion Chapati

Chapati is an Indian flatbread similar in size to pita bread, but with a firmer texture so it holds up better for a sandwich with a moist filling. You can also use a tortilla, lavash, or whole wheat wrap, but check the calories; more than 100 and the starch count is 2. If you want to fill a pita pocket, make sure the vegetables are slightly cooled or the bread will not hold together. And to make a pita sandwich ahead, pack the vegetables and pita in separate containers or the bread will get soggy and fall apart.

1 tablespoon extra virgin olive oil
½ cup sliced red onion (½ small)
Salt
1 cup sliced fresh or frozen green and/or red bell pepper
   (1 medium fresh)
2 tablespoons shredded cheese (Cheddar, mozzarella, Swiss, etc.)
1 (7-inch) whole wheat chapati

Heat the olive oil in a small skillet over medium-low heat. Add the onion, season lightly with salt, and cook for 5 minutes. Add the bell pepper and season lightly with salt. Cook until the vegetables are soft and have absorbed most of the oil, about 10 minutes. If you have used frozen peppers, make sure any excess water has evaporated.

Sprinkle the cheese over the vegetables and cook until it melts, about 3 minutes. Covering the pan with a lid will help the cheese melt quickly. Place the cooked vegetables in the center of the chapati and roll up. If you are not eating the sandwich immediately, wrap tightly in foil.

MAKES I SERVING.
CALORIES: 360; STARCH: I (chapati); VEGETABLES: 3 (I onion, 2 pepper); FAT: I (olive oil);
FRUIT: 0; DAIRY: ½ (cheese)

Cheese and Broccoli Wrap

We used a round flatbread (wrap), but you can use a rectangular one such as a lavash *and roll it up jelly-roll style instead of folding it.*

1 tablespoon extra virgin olive oil
1 cup fresh or thawed frozen chopped broccoli
Salt and black pepper
¼ cup (1 ounce) shredded Cheddar or mozzarella cheese
1 (7-inch) whole wheat wrap

Heat the olive oil in a small skillet over medium heat. Stir in the broccoli and season with salt and pepper. Reduce the heat to medium-low and cook until the broccoli is tender, 10 to 15 minutes. Sprinkle the cheese over the broccoli and cook until it melts, 2 to 3 minutes. Covering the pan with a lid will help the cheese melt quickly.

Place the wrap on a flat surface. Use a rubber spatula to spread the broccoli and cheese into a circle in the center of the wrap, leaving about a 1-inch border. Fold the bottom of the wrap over the filling and hold it there with your thumbs while your fingers fold in the two sides. Then roll up, leaving the top edge on the bottom. Give the pocket some gentle squeezes so it is encouraged to hold together. You can also use a tooth-pick to secure the top edge. If you are packing the sandwich to go, wrap tightly in aluminum foil.

VARIATIONS: This is a simple sandwich filling and there are a number of ways to jazz it up: Cook about 10 chopped black olives with the broccoli (add ½ fat) and use mozzarella cheese—or better yet, provolone; scatter some roasted red bell peppers or chopped sun-dried tomatoes on top of the filling before rolling it up; sauté some chopped garlic with the broccoli.

MAKES 1 SERVING.
CALORIES: 430; STARCH: 2 (wrap); VEGETABLES: 2 (broccoli); FAT: 1 (olive oil); FRUIT: 0; DAIRY: 1 (cheese)

Grilled Cheese and Spinach Sandwich

When is a grilled cheese sandwich the best that it can be? When it is cooked in olive oil. Use the smallest pan that will hold the sandwich and make sure the oil is hot before adding the bread or it will stick to the pan. This sandwich travels well so it can be made in the morning, then wrapped in foil or put in a plastic container to take along for lunch. Enjoy at room temperature or heat briefly in a microwave.

2 tablespoons extra virgin olive oil
2 cups baby spinach or ⅓ cup thawed frozen spinach
Salt
2 slices whole wheat bread
1 ounce (¼ cup) shredded cheese (Cheddar, Muenster, mozzarella,
    or Swiss)

Heat 1 tablespoon of the olive oil in a small pan over medium-low heat. Add the spinach, season with salt, and cook until wilted; frozen spinach should be cooked for 10 minutes so any water evaporates and the spinach absorbs the oil.

Place one slice of bread on a plate or cutting board. Put the cheese on the bread. Use a fork to remove the spinach from the pan and put it on top of the cheese. Cover with the second slice of bread.

Still with the same pan, heat the remaining 1 tablespoon olive oil over medium heat. Put the sandwich in the pan and weigh it down. An easy way to do this is with a small plate and a heavy can on top of it. Cook until the bottom slice of bread is browned, 3 to 4 minutes. Use a spatula to turn the sandwich, return the weight, and cook for 3 to 4 minutes, until the second side is browned.

VARIATIONS: You can use any vegetable in place of the spinach. Grilled cheese and tomatoes or roasted red bell peppers are great combinations. For a popular combination of flavors, use mozzarella cheese and fry a few slices of tomatoes in the oil until they are lightly colored, then cover them with several leaves of fresh basil.

MAKES I SERVING
CALORIES: 530; STARCH: 2 (bread); VEGETABLES: 2 (spinach); FAT: 2 (olive oil); FRUIT: 0;
DAIRY: I (cheese)

Portobello Mushroom, Onion, and Roasted Red Peppers with Provolone on a Whole Wheat Roll

Portobello mushrooms have a concentrated rich flavor and a dense meaty texture, making them perfect for sandwiches. I "construct" this sandwich by keeping the vegetables large and cooking them in the same pan, but separately, then layering them on a roll. I suppose I could have sliced all the vegetables and cooked them together, but the finished sandwich would not have been half so pretty. If you are taking this to work, pack the vegetables and bread separately so the bread does not get soggy. You can also use a sandwich round, in which case the starch count will be 1.

2 tablespoons extra virgin olive oil
¼-inch slice red onion in one piece
Salt
¼ cup roasted red bell pepper cut in ¼-inch-wide strips (see page 220)
    or sliced jarred roasted peppers
1 (3-ounce) portobello mushroom cap (approximately 4 inches diameter)
1 ounce provolone cheese, sliced or shredded (¼ cup)
1 (2-ounce) whole wheat roll

Heat the olive oil in a pan large enough to hold the onion slice, mushroom, and pepper separately. Place the onion ring in the pan, keeping it as a slice. Season with salt. Cook until slightly browned on the bottom, about 5 minutes. Turn the onion slice over carefully so the rings stay together, and put the peppers on one side of the pan and the mushroom on the other. Season with salt. Cook, stirring the peppers and turning over the mushroom a few times, until almost all of the oil has been absorbed, 8 to 10 minutes. Put the cheese on top of the mushroom and cook until it melts, 2 to 3 minutes.

Layer the vegetables on the roll in this order: the onion slice, the mushroom with the cheese, and the pepper strips on top.

MAKES 1 SERVING
CALORIES: 500; STARCH: 2 (roll); VEGETABLES: 2 (½ each red onion and pepper,
1 mushroom); FAT: 2 (olive oil); FRUIT: 0; DAIRY: 1 (cheese)

Roasted Eggplant Sandwich

Roasted vegetables make great sandwiches, and if you have containers of them in the refrigerator and freezer, they make great "quick" sandwiches. Refrigerated vegetables will reheat very quickly. You don't have to thaw frozen vegetables; just pop them into a warm pan and cook over medium-low heat, turning often, until they thaw and heat up. This sandwich uses eggplant, but you can substitute any roasted vegetable you like.

1 (1½-inch-thick) slice roasted eggplant (3 to 4 inches in diameter), skin
   on (see page 217)
1 whole wheat sandwich round (see page 82)
¼ cup (1 ounce) shredded part skim, low-moisture mozzarella cheese
1 cup arugula

Heat a small pan over medium-low heat and slip in the eggplant. Cook, turning often, until it is warm throughout. Toast the sandwich round while the eggplant is heating.

Layer the eggplant, cheese, and then the arugula on the sandwich round.

VARIATIONS: Use mild Cheddar cheese and spread a toasted whole wheat sandwich round with 2 tablespoons Roasted Red Pepper Spread/Dip (page 131) before layering the other ingredients; add ½ starch and ½ fat. Or, dice the eggplant, spread a wrap with 2 tablespoons Roasted Red Pepper Spread/Dip (page 131), substitute feta for the mozzarella, and scatter ¾ cup fresh baby spinach over the ingredients; add ½ starch, ½ fat, and 1 vegetable.

MAKES 1 SERVING
CALORIES: 315; STARCH: 1 (sandwich round); VEGETABLE: 2 (1 each eggplant and arugula);
FAT: 1 (olive oil from the roasted eggplant); FRUIT: 0; DAIRY: 1 (cheese)

Bean, Spinach, and
Roasted Red Pepper Sandwich

It's not so long ago that Saturday night supper in New England meant a large crockery pot of homemade baked beans—and sometime in the next few days, cold bean sandwiches. This modern, healthy version is as satisfying as were those baked bean sandwiches, and you don't have to own a crockery bean pot.

2 tablespoons extra virgin olive oil
½ cup roasted red bell pepper strips, ¼-inch wide (see page 220) or
   jarred red peppers
Salt and black pepper
2 cups baby spinach or ⅓ cup thawed frozen spinach
1 whole wheat sandwich round (see page 82)
½ cup cooked cannellini beans (see page 128), or canned beans, drained
   and rinsed

Heat the olive oil in a small pan over medium-low heat. Stir in the bell peppers, season with salt and black pepper, and cook until warmed, about 5 minutes. Add the spinach, season lightly with salt, and cook until wilted; frozen spinach should be cooked for 10 minutes so it absorbs the oil.

Toast the sandwich round while the vegetables are cooking.

Add the beans to the vegetables, season with black pepper, and cook until heated through, 3 to 4 minutes. Spoon the mixture onto the bread and enjoy.

MAKES I SERVING
CALORIES: 490; STARCH: 2 (I beans, I bread); VEGETABLES: 3 (I pepper, 2 spinach);
FAT: 2 (olive oil); FRUIT: 0; DAIRY: 0

Spinach, Onion, and Mushroom Sandwich

Want a sandwich in a real big hurry? This is it. We like it on a sandwich round, but you can use any bread you like.

1 tablespoons extra virgin olive oil
½ cup chopped red onion (½ small)
Salt
½ cup sliced mushrooms
2 cups baby spinach or ⅓ cup thawed frozen chopped spinach
1 whole wheat sandwich round (see page 82)

Heat the olive oil over medium heat in a skillet large enough to accommodate the spinach. Stir in the onion, season with salt, and cook until it begins to soften, about 10 minutes. Add the mushrooms, season lightly with salt, and cook until their juices run. Toss in the spinach, season lightly, and cook until the spinach wilts and the vegetables have absorbed almost all the oil. Spoon the mixture onto the sandwich round and enjoy.

MAKES I SERVING.
CALORIES: 310; STARCH: 1 (bread); VEGETABLES: 4 (1 each onion and mushrooms, 2 spinach); FAT: 1 (olive oil); FRUIT: 0; DAIRY: 0

Peanut Butter and Fruit Spread Sandwich

I doubt that you really need a recipe for a peanut butter and jelly sandwich. But I have it here because you may be searching for something you can do in a hurry when you don't want to cook, and that won't throw off your diet. So the recipe is simple—but with rules: Use whole wheat bread, no more than 2 tablespoons of peanut butter (one with no trans fats or partially hydrogenated oil) and instead of jelly or jam use a fruit spread (one that is 100 percent fruit). Try some additional fruit on the sandwich, if you like: 2 tablespoons raisins or dried cranberries, or ½ cup banana slices; add 1 fruit. I find it easier to measure peanut butter by weight, so I have given you those directions.

2 slices whole wheat bread

2 tablespoons (32 grams) peanut butter

1 tablespoon fruit spread (100 percent fruit)

Put a piece of bread on a scale, zero it out and spoon on enough peanut butter to measure 32 grams. Spread the peanut butter on the bread. Spread the fruit spread on the other piece of bread. Put the slices together.

MAKES I SERVING

CALORIES: 385; STARCH: 2 (bread); VEGETABLES: 0; FAT: 2 (peanut butter); FRUIT: ¼ ; DAIRY: 0

Veggie Burger Sandwich

You can make your own veggie burgers (page 135) or purchase them in a supermarket. They are sold frozen and either in the health food sections or with other frozen premade foods. But do try making your own. They're very easy and you can flavor them any way you want. The veggie burgers you buy in the market vary in calories, usually from 70 to 200 calories per patty. Ours have 275 calories, are larger than the frozen varieties, and most importantly, are made with the all-important extra virgin olive oil. So, if you choose a frozen burger, be sure and use olive oil to cook the burger or elsewhere in the meal. If you want cheese on your burger, cover the burger with a 1-ounce slice of any variety you like right after you flip the patty in the pan; add 1 dairy.

1 tablespoon extra virgin olive oil

¼ cup sliced red onion

¼ cup green and/or red bell pepper strips, frozen or fresh

Salt and black pepper

1 Black Bean Veggie Burger (page 135)

1 whole wheat sandwich round (see page 82)

Heat the olive oil in a small pan over medium heat. Add the onion and pepper strips and season with salt and black pepper. Cook until the vegetables begin to soften, about 10 minutes, less time if you want your vegetables to remain firmer.

Push the vegetables to the side and put the veggie burger in the pan. Cook, undisturbed, until browned, about 5 minutes. Turn it over carefully with a spatula and move the vegetables on top. Cover and cook for 3 to 4 minutes, until the burger is browned and heated through. Transfer the veggie burger and vegetables to the bottom half of the sandwich round, cover with the top half, and eat while hot.

MAKES I SERVING
CALORIES: 515; STARCH: 2 (1 veggie burger, 1 sandwich round) ; VEGETABLES: approximately 2 (1 each onion and peppers); FAT: 1⅓ (olive oil); FRUIT: 0; DAIRY: ½ (cheese used to make the burger)

Pita Chips

Not a sandwich but something good to have on hand to accompany soup, eat alone, or use for dips. Make several batches and store in an airtight container (so they don't lose their crispness). Enjoy warm or at room temperature. These are simply salted, but you can sprinkle them with any herb or spice you like.

1 small (6½-inch) whole wheat pita
1 tablespoon extra virgin olive oil
Salt and black pepper

Preheat the oven to 375°F.

Cut the pita into 4 pie-shaped pieces. Separate each piece at the fold so you have 8 chips. Brush the inner sides with olive oil and season with salt and pepper. Place on a baking sheet in a single layer, oil side up. Bake the chips for about 10 minutes, or until browned.

MAKES I SERVING, 8 CHIPS
CALORIES: 290 (35 calories for each chip); STARCH: 2 (pita); VEGETABLES: 0; FAT: 1 (olive oil); FRUIT: 0; DAIRY: 0

Black Bean, Corn, and Tomato Salad

This salad is a snap to make and a good one to take on the road since there are no greens that would wilt. A tablespoon or two of chopped fresh cilantro is a nice addition.

½ cup cooked black beans (see page 128), or canned beans, drained and
    rinsed
½ cup canned corn, drained, or thawed frozen corn
½ cup chopped red and/or green bell pepper (½ medium)
½ cup diced tomatoes (½ medium; about 3 ounces)
¼ cup chopped red onion
2 tablespoons extra virgin olive oil
2 teaspoons lime juice
Salt and black pepper

Toss the beans, corn, bell pepper, tomatoes, and onion together in a bowl. Whisk the olive oil and lime juice together and season with salt and pepper. Stir the dressing into the beans and vegetables.

MAKES I SERVING
CALORIES: 450; STARCH: I (beans); VEGETABLES: 3½ (½ onion, I each corn, peppers, and tomatoes); FAT: 2 (olive oil); FRUIT: 0; DAIRY: I

GOOD TO KNOW

Dressing Salads

If you eat salads often and want the convenience of a dressing already made, increase the amounts, proportionally, of any of the dressings in the salad recipes. The dressings in those recipes have a higher percentage of acid (lemon juice or vinegar) to oil than the more common three or four parts oil to one part acid. This saves calories. Even a simple, low-calorie salad can become high caloric when a dressing is added, so play with the flavors until you find a balance you like. Use any type of vinegar for the acid—red or white wine vinegar, balsamic vinegar, cider. To vary the flavor, add some chopped garlic or shallots, a dash of Dijon mustard, and/or some fresh herbs.

Wheat Berry Salad

Wheat berries are whole, unprocessed wheat kernels with the bran and germ intact so they have all the nutrients that wheat has to offer. If you can't find them in your grocery store, try a natural food store. They are worth the trip. There are a number of varieties of wheat berries—spelt, farro, kamut, to name a few—but most common are the hard, red winter wheat berries used here. If you don't soak the wheat berries overnight, you'll need to cook them longer, about 1 hour and 10 minutes.

½ cup hard red winter wheat berries
Salt and black pepper
1 cup chopped pecans
1 cup dried cranberries or dried cherries
1 cup chopped red onion (½ medium)
½ cup extra virgin olive oil
¼ cup balsamic vinegar

Rinse the wheat berries in a colander with cold water. Transfer to a large pot and add water to cover by 2 inches. Cover and let soak overnight.

Bring at least 4 cups of water to a boil and add ½ teaspoon salt. Drain the wheat berries and stir into the pot. Reduce the heat to a slow boil and cook, uncovered, until the wheat is tender but still chewy, about 50 minutes. You can cook them less if you want the grain to be firmer. Drain off any excess water and let cool.

Mix the wheat berries, pecans, dried fruit, and onion together in a large bowl. Whisk together the olive oil, vinegar, and salt and pepper. Pour the dressing over the salad, mix together, and refrigerate until ready to serve. Serve cold or at room temperature.

MAKES 8 SERVINGS, ABOUT ¾ CUP EACH
CALORIES: 285; STARCH: 1 (wheat berries); VEGETABLES: less than ½ ; FAT: 2 (1 olive oil, 1 pecans); FRUIT: 1 (dried fruit); DAIRY: 0

Barley Salad

The barley should be cooled for this salad so make it well ahead—even a few days—and let it chill in the refrigerator. The salad itself is best made a few hours before serving so the flavors will blend. You can make it the day before, but add the arugula close to serving or it will wilt.

2 cups cooked barley (see page 138)
2 cups diced red, green, and/or yellow bell peppers (2 medium)
1 cup chopped red onion (½ medium)
2 cups arugula, ripped into small pieces
½ cup extra virgin olive oil
¼ cup balsamic vinegar
Salt and black pepper

Put the barley in a large bowl. If it is sticky, use a fork to separate and fluff the grains. Add the bell peppers, onion, and arugula and toss well so they are evenly distributed. Whisk together the olive oil and vinegar in a separate bowl and season with salt and pepper. Drizzle over the salad.

MAKES 8 SERVINGS, ABOUT I CUP EACH
CALORIES: 375; STARCH: 2 (barley); VEGETABLES: 2 (½ each onion and arugula, I peppers); FAT: I (olive oil); FRUIT: 0; DAIRY: 0

DECREASING THE RECIPE
The yield for this salad is large because I found that it was one of the recipes the women in the study especially liked to serve to families and company. You can make half the recipe, or a quarter. The count for each serving remains the same.

Tabbouleh

Bulgur wheat (bulghur), a whole grain food, is wheat kernels that have been steamed, dried, and cracked. With a pleasantly chewy texture it is perfect for salads. Tabbouleh (tabouli) is a Middle Eastern salad overachiever that has built its reputation on bulgur. Although bulgur can be boiled (see page 138), the texture is better for salads if it is just given a bath in boiling water, as below.

4 ounces (¾ cup) bulgur wheat
Salt and black pepper
¼ cup slivered almonds
6 tablespoons extra virgin olive oil
2 tablespoons lemon juice
2 cups cherry or grape tomatoes, cut in half (about 10 ounces)
2 cups fresh flat-leaf parsley, chopped
1 cup thinly sliced scallions (3 or 4)
¼ cup fresh mint leaves, torn into small pieces

Put the bulgur in a bowl and stir in 2 cups boiling water and ½ teaspoon salt. Cover with an inverted plate and let stand until the bulgur is tender and most of the water is absorbed, 20 to 30 minutes. Drain the bulgur in a sieve, pressing gently to remove excess water, and then return to the bowl.

While the bulgur is soaking, toast the almonds: Preheat the oven to 350°F. Arrange the almonds on a baking sheet in a single layer. Toast in the oven for 4 to 5 minutes, until lightly browned. Let cool.

Whisk together the olive oil, lemon juice, and salt and pepper in a large bowl. Add the tomatoes, parsley, scallion, and mint. Stir in the bulgur and almonds. Let the salad cool before serving.

MAKES 4 SERVINGS, ABOUT 2½ CUPS EACH
CALORIES: 360; STARCH: 1 (bulgur); VEGETABLES: 2½ (½ onion, 1 each tomato and parsley); FAT: 2 (½ almonds, 1½ olive oil); FRUIT: 0; DAIRY: 0

Bulgur and Summer Vegetable Salad

A nice variation on tabbouleh.

4 ounces (¾ cup) bulgur wheat
Salt and black pepper
2 cups cherry or grape tomatoes, sliced in half (about 10 ounces)
2 cups chopped unpeeled cucumber (1 medium)
1 cup diced red onion (½ medium)
4 cups baby spinach, torn into pieces
6 tablespoons extra virgin olive oil
2 tablespoons lemon juice

Put the bulgur in a bowl and stir in 2 cups boiling water and ½ teaspoon salt. Cover with an inverted plate and let stand until the bulgur is tender and most of the water is absorbed, 20 to 30 minutes. Drain the bulgur in a sieve, pressing gently to remove excess water, and then return it to the bowl.

Add the tomatoes, cucumber, onion, and spinach to the bulgur and distribute as evenly as possible. Whisk together the olive oil, lemon juice, and salt and pepper. Pour over the salad and toss together. Cool before serving.

MAKES 4 SERVINGS, ABOUT 3 CUPS EACH
CALORIES: 315; STARCH: 1; VEGETABLES: 3½ (½ onion, 1 each tomato, cucumber, and spinach); FAT: 1½ (olive oil); FRUIT: 0; DAIRY: 0

Hearty Vegetable Salad

This is a vegetable-generous salad that travels well if you use a sturdy lettuce such as iceberg or romaine. You could also use Bibb or Boston lettuce, but then pack the lettuce separately for travel and join it with the vegetables when you are ready to eat it. This is a nice place to use your very best extra virgin olive oil. If you want to skip the nuts, subtract 100 calories and 2 fat servings.

2 cups lettuce, torn into bite-sized pieces
½ cup cooked cannellini, black, or kidney beans (see page 128), or
    canned beans, drained and rinsed
½ cup thinly sliced carrot (1 medium)
½ cup chopped red or green bell pepper (½ medium)
1 cup grape or cherry tomatoes (about 5 ounces)
2 tablespoons walnut pieces
2 tablespoons raisins
1 tablespoon extra virgin olive oil
2 teaspoons balsamic vinegar
Salt and black pepper

Mix the lettuce, beans, carrot, bell pepper, tomatoes, walnuts, and
raisins in a bowl. Whisk together the olive oil and vinegar and season
with salt and pepper. Pour the dressing over the salad and toss well.

MAKES 1 SERVING
CALORIES: 480; STARCH: 1 (beans); VEGETABLES: 6 (1 each carrots and peppers, 2 each
tomatoes and lettuce); FAT: 3 (1 olive oil, 2 nuts); FRUIT: 1 (raisins); DAIRY: 0

Greek Salad

*A popular salad that makes an easy lunch. You can serve it as it is, or spoon it on top of
romaine lettuce leaves. It also makes a great sandwich when tucked inside a wrap;
count the starch and calories for the wrap. Cut grape tomatoes in half if they are large.*

2 tablespoons extra virgin olive oil
2 teaspoons red wine vinegar
2 teaspoons lemon juice
½ teaspoon dried oregano
Salt and black pepper
1 cup grape tomatoes, or 2 plum tomatoes (about 5 ounces), quartered
½ cup sliced unpeeled cucumber (½ small)
½ cup sliced green bell pepper (½ medium)
6 pitted large black olives (about 1 ounce)
¼ cup (1 ounce) feta cheese

Whisk together the olive oil, vinegar, lemon juice, oregano, and salt and pepper. Mix together the tomatoes, cucumber, pepper, olives, and feta in a large bowl. Pour the dressing over the salad and toss together.

MAKES I SERVING
CALORIES: 420; STARCH: 0; VEGETABLES: 4 (1 each cucumbers and peppers, 2 tomatoes); FAT: 2½ (½ olives, 2 olive oil); FRUIT: 0; DAIRY: 1 (cheese)

Butter Bean Salad

Butter beans are the large variety of beans also known as Fordhook lima beans. They are not grown-up baby lima beans even though they resemble them—pale, plump, and kidney-shaped. Butter beans are fuller flavored than small lima beans and make a great base for a salad.

2 cups canned butter beans, drained and rinsed
2 cups chopped unpeeled cucumber (1 medium)
2 cups cherry or grape tomatoes, cut in half (about 10 ounces)
2 cups canned corn, drained, or thawed frozen corn
1 cup diced red onion (½ medium)
½ cup cilantro leaves
½ cup extra virgin olive
¼ cup lime juice
½ teaspoon ground cumin
¼ teaspoon ground cayenne pepper
½ teaspoon salt

Combine the beans, cucumber, tomatoes, corn, and onion in a bowl. Whisk together the cilantro, olive oil, lime juice, cumin, cayenne, and salt in another bowl. Pour the dressing over the vegetables and beans and mix together. Cover and refrigerate for at least 2 hours before serving so the flavors will blend.

MAKES 4 SERVINGS, ABOUT 2½ CUPS EACH
CALORIES: 475; STARCH: 1 (beans); VEGETABLES: 3½ (½ onion, 1 each cucumber, tomatoes, and corn); FAT: 2 (olive oil); FRUIT: 0; DAIRY: 0

Salade Niçoise

This classic French salad is particularly nice in the summer when made with new pota-toes, fresh green beans, and ripe garden tomatoes. If the green beans are tender and young, I like them raw, but you can blanch them (cook in boiling, salted water) for 3 to 5 minutes until they are tender but still crisp. Drain immediately and plunge into a bowl of ice water to stop the cooking and to retain the bright green color. Then drain again and wrap in a kitchen towel to dry. The potatoes can be slightly warm or completely cool when you make the salad.

3 ounces unpeeled red potatoes, scrubbed clean and cut into 1-inch
     pieces
Salt and black pepper
1 cup fresh green beans, cut into 2-inch pieces (about 4 ounces)
½ cup chopped red onion (½ small)
1 small ripe tomato, cut into eighths, or ½ cup cherry or grape tomatoes,
     halved (2 to 3 ounces)
1 tablespoon extra virgin olive oil
2 teaspoons red wine vinegar
1 teaspoon Dijon mustard
Several lettuce leaves, preferably Boston or Bibb
2 ounces drained canned or packaged tuna packed in water
1 large egg, hard boiled, shell removed, and quartered
10 pitted small black olives (about 1 ounce)
1 tablespoon capers, rinsed and drained

Put the potatoes in a small sauce pan. Add cold water to cover by 1 inch and ½ teaspoon salt. Cover, bring to a boil, and reduce the heat to medium-low. Simmer until the potatoes are tender when pierced with a fork, 5 to 7 minutes. Drain and let cool briefly.

Combine the potatoes, green beans, tomatoes, and onions in a bowl. Whisk together the olive oil, vinegar, mustard, and salt and pepper. Pour over the vegetables and toss together well.

Lay the lettuce leaves on a plate and put the tuna in the center. Sur-round with the vegetables. Make a nice arrangement with the egg quar-ters and olives and scatter the capers on top of the tuna.

MAKES I SERVING
CALORIES: 450; STARCH: I (potatoes); VEGETABLES: 4 (I each onion and tomato, 2 green beans); FAT: 1½ (½ olives, I olive oil); FRUIT: 0; DAIRY: I (egg); SEAFOOD: 2 ounces (tuna)

Tuna and Bean Salad

Just ask any Italian and they will tell you that tuna and cannellini beans make a great marriage. This healthy version is proof in point.

2 cups salad greens
3 ounces canned tuna packed in water, drained
½ cup cooked cannellini beans (see page 128), or canned beans, drained and rinsed
½ cup grape or cherry tomatoes, cut in half
1 tablespoon capers, rinsed and drained
2 tablespoons extra virgin olive oil
2 teaspoons balsamic vinegar
Salt and black pepper

Arrange the salad greens on a plate. Top with the tuna, beans, tomatoes, and capers. Whisk together the olive oil, vinegar, and salt and pepper and pour over the salad.

MAKES I SERVING
CALORIES: 510; STARCH: I (beans); VEGETABLES: 3 (I tomatoes, 2 lettuce); FAT: 2 (olive oil); FRUIT: 0; DAIRY: 0; SEAFOOD: 3 ounces

White Bean and Spinach Salad

Here's an easy summer lunch that is good at room temperature or chilled.

2 cups baby spinach
1 cup cooked cannellini beans (see page 128), or canned beans, drained
   rinsed
1 cup cherry or grape tomatoes (about 5 ounces)
1 tablespoon extra virgin olive oil
2 teaspoons balsamic vinegar
Salt and black pepper

Combine the spinach, beans, and tomatoes in a bowl. Stir in the olive oil
and vinegar. Season with salt and pepper and gently stir to combine.

MAKES I SERVING
CALORIES: 390; STARCH: 2 (beans); VEGETABLES: 4 (2 tomatoes, 2 spinach); FAT: I (olive
oil); FRUIT: 0; DAIRY: 0

Strawberry and Spinach Salad

*This is a really nice "company" salad. Just multiply all the ingredients by the number of
people you are serving. Do not add the strawberries to the spinach until ready to serve
or the color will bleed onto the greens.*

1 cup strawberries, fresh or thawed frozen, sliced
1 tablespoon extra virgin olive oil
2 teaspoons balsamic vinegar
1 small garlic clove, minced
Salt and black pepper
2 cups baby spinach

Place the strawberries in a bowl. Whisk together the olive oil and vinegar. Add the garlic and salt and pepper. Pour over the strawberries,
cover, and marinate for 2 to 3 hours.

When ready to serve, arrange the spinach on a plate. Top with the marinated strawberries.

MAKES I SERVING
CALORIES: 180; STARCH: 0; VEGETABLES: 2 (spinach); FAT: I (olive oil); FRUIT: I (strawberries); DAIRY: 0

Pear and Cranberry Salad

This is a nice salad to keep in mind for the holidays. It is easy to increase for a large gathering of family or friends. If you are making it ahead of time, squeeze some lemon juice on the pear slices to prevent them from browning.

2 cups mixed salad greens
8 grapes, cut in half lengthwise
2 tablespoons dried cranberries
1 tablespoon pecan halves
½ medium pear (any variety), cut into thin slices (4 ounces)
1 tablespoon extra virgin olive oil
2 teaspoons balsamic vinegar
Salt and black pepper

Spread the salad greens on a plate. Scatter the grape halves, cranberries, and nuts over the lettuce and arrange the pear slices on top.

Whisk together the olive oil, vinegar, and salt and pepper. Drizzle over the salad.

MAKES I SERVING
CALORIES: 340; STARCH: 0; VEGETABLES: 2 (salad greens); FAT: 2 (I each olive oil and nuts); FRUIT: 2 (½ grapes, I each pear and dried cranberries); DAIRY: 0

Tomato-Pepper Pasta Salad

If you are making pasta for dinner, make some extra for this salad. Three ounces dried pasta is approximately 2 cups cooked. Chopped fresh parsley or whole basil leaves are good in this salad.

3 ounces whole wheat pasta, cooked, drained, and cooled
1 cup cherry or grape tomatoes, cut in half (about 5 ounces)
½ cup sliced green and/or red bell pepper (½ medium)
1 cup arugula or baby spinach
2 tablespoons extra virgin olive oil
Salt and black pepper

Mix the pasta, tomatoes, bell pepper, and arugula together in a bowl. Drizzle with the olive oil, season with salt and pepper, and toss together.

MAKES 1 SERVING
CALORIES: 550; STARCH: 3 (pasta); VEGETABLES: 4 (1 each pepper and arugula, 2 tomatoes); FAT: 2 (olive oil); FRUIT: 0; DAIRY: 0

Basic Potato Salad

Potato salad is a great vehicle for any number of additions. Try scallions, celery, parsley, snipped fresh chives, and a splash of vinegar. Or give it a German twist with chopped dill pickle, celery leaves, a dash of cider vinegar, and a little paprika. French style? A teaspoon or two of Dijon mustard, some chopped fresh tarragon, a few capers, and red wine vinegar.

9 ounces unpeeled red-skinned potatoes, cut into bite-size pieces
Salt and black pepper
1 cup diced bell pepper (1 medium)
½ cup chopped red onion (½ small)
1 tablespoon extra virgin olive oil

Put the potatoes in a small saucepan and add cold water to cover by 1 inch and ½ teaspoon salt. Cover and bring to a boil. Reduce the heat to medium-low and simmer until the potatoes are tender when pierced with a fork, 5 to 7 minutes. Drain and let cool briefly.

Mix the potatoes, bell pepper, and onion together in a bowl. Drizzle with the olive oil, season with salt and pepper, and stir together.

MAKES 1 SERVING
CALORIES: 510; STARCH: 3 (potatoes); VEGETABLES: 3 (1 onion, 2 pepper); FAT: 1 (olive oil); FRUIT: 0; DAIRY: 0

Roasted Beet, Walnut, and Gorgonzola Salad

This salad is good cold, at room temperature, or, if you have just roasted the beets, warm. It is a most satisfying combination of ingredients.

1 cup roasted beets cut in 1-inch cubes (see page 216)
1 tablespoon walnut pieces
½ ounce Gorgonzola cheese (1 tablespoon)
½ medium pear (any variety), chopped (4 ounces)
2 teaspoons extra virgin olive oil
2 teaspoons lemon juice
Salt and black pepper
2 cups arugula

Combine the beets, walnuts, Gorgonzola, and pear in a bowl. Whisk the olive oil, lemon juice, and salt and pepper together and pour over the salad. Spread the arugula out on a plate and top with the salad.

MAKES 1 SERVING
CALORIES: 395; STARCH: 0; VEGETABLES: 2 (beets); FAT: 2 ⅓ (⅓ dressing, 1 walnuts, 1 roasted beets); FRUIT: 1 (pear); DAIRY: ½ (cheese)

Roasted Tomato and
Cannellini Bean Salad

This salad is best if you start with garden ripe tomatoes. I particularly like some of the heirloom varieties that are becoming increasingly available. Make the salad ahead because its flavors are enhanced by marinating. Enjoy with a piece of bread.

1 large tomato, quartered (about 10 ounces)
2 tablespoons extra virgin olive oil
Salt and black pepper
½ cup cooked cannellini beans (see page 128), or canned beans, drained
    and rinsed
½ cup red onion slices (½ small)
1 garlic clove, minced
6 large pitted olives (about 1 ounce), cut in half
¼ cup chopped fresh flat-leaf parsley
2 fresh basil leaves, torn into small pieces

Preheat the oven to 425°F. Place the tomatoes in a roasting pan, drizzle with the olive oil, and season with salt and pepper. Roast for 30 to 40 minutes, until the tomatoes are slightly wrinkled. Let cool in the pan.

Transfer the tomatoes with any oil from the pan to a serving bowl. Add the beans, onion, garlic, olives, parsley, and basil and stir to combine. This salad tastes best if allowed to marinate for about 1 hour.

MAKES I SERVING
CALORIES: 475; STARCH: I (beans); VEGETABLES: 5 (I onion, 4 tomatoes);
FAT: 2½ (½ olives, 2 olive oil); FRUIT: 0; DAIRY: 0

Wild Rice Salad

Wild rice isn't actually rice at all. It's a whole grain marsh grass prized for its nutty flavor and chewy texture. This recipe makes either a salad serving for 4 or a side dish for 8, in which case you should divide the allowances in half.

6 tablespoons extra virgin olive oil
3 tablespoons balsamic vinegar
Salt and black pepper
1 cup chopped fresh flat-leaf parsley
1 cup chopped red onion (½ medium)
½ cup dried cranberries or dried cherries
6 tablespoons wild rice (2 ounces)
5 tablespoons brown rice (2 ounces)
¼ cup pecan halves

Whisk together the olive oil, balsamic vinegar, and salt and pepper. Add the parsley, onion, and dried fruit. Let sit for about 1 hour.

Cook the wild rice and brown rice together as directed on page 138. Chill in the refrigerator.

Preheat the oven to 350°F. Arrange the pecans on a baking sheet in a single layer and toast in the oven for 4 to 5 minutes, until lightly toasted. Keep your eye on them because any nut can burn very quickly.

Combine the cooled rice and toasted pecans in a large bowl. Pour the dressing over the salad and gently combine.

MAKES 4 SERVINGS, 1¼ CUPS EACH
CALORIES: 410; STARCH: 1 (rice); VEGETABLES: 1 (½ each onion and parsley);
FAT: 2½ (1 pecans, 1½ olive oil); FRUIT: 1 (dried cranberries or cherries)

Potatoes, Beans, and Grains

Potatoes

Potatoes have gone through some rough times. Dieters shunned them as fattening. They became the villains of the low-carb diet fad, the baddies of the nonsensical avoid-all-white-carbohydrates diet. Not since the Great Famine of Ireland has the humble tuber seen such woes. And those woes are uncalled for. Potatoes are not fattening—it's the half stick of butter and cup of sour cream on top that is fattening. And when deep fried in lard they'll pack on the pounds. But the potato has nothing to do with all those calories. One small 6-ounce potato has only 120 calories (20 calories per ounce) and uses up just 2 of your daily starches. Potatoes are "good carbs." They have fiber, vitamins B_6 and C, lots of potassium, and phytonutrients that help keep your blood pressure normal. All for about twenty-five cents a serving. Wow! All varieties of potatoes are allowed—baking (russet or Idaho), new potatoes, red potatoes, purple potatoes, Yukon Golds, fingerlings, round whites (often called boiling potatoes), and sweet potatoes and yams. Because the skins have fiber and phytonutrients as well as flavor, I always leave them on, just giving the spuds a good scrubbing under running water before cooking.

Be watchful when buying baking potatoes. They vary in size from 6 to 14 ounces. If you are buying loose baking potatoes, search through the bin for the smallest ones you can find and weigh them at the grocery store; produce sections commonly have scales. The average baking potato sold loose weighs 12 ounces, or, put another way, uses up 4 of your 6 daily allowance of starch. As a general rule, potatoes sold in 2- or 5-pound bags tend to be smaller, usually 6 to 8 ounces.

The potato recipes that follow here are main courses. You will find more potato recipes in Chapter 10 on side dishes.

Mashed Potato Dinner

This is total comfort food for a winter dinner. Try a variety of leafy greens in place of the spinach. This reheats nicely in the microwave, so if you want to take it to work, make it the night before and store in a microwaveable container.

2 tablespoons extra virgin olive oil
2 cups baby spinach or ⅓ cup thawed frozen chopped spinach
Salt and black pepper
¼ cup chopped roasted red bell pepper (see page 220) or ¼ cup jarred
    red pepper
½ cup cooked cannellini beans (see page 128), or canned beans, drained
    and rinsed
6 ounces unpeeled red potatoes, scrubbed clean and coarsely chopped
¼ cup nonfat or 1% milk, warmed

Heat the olive oil in a medium pan over medium heat. Add the spinach, season with salt and pepper, and cook until it wilts, 1 or 2 minutes; frozen spinach should be cooked for 7 minutes, or until any excess water has evaporated. Add the red peppers, stir to coat with oil, and cook until they are warm, 3 or 4 minutes. Stir in the beans and keep warm.

Meanwhile, put the potatoes in a small saucepan and cover with cold water by 1 inch. Add about ½ teaspoon salt, cover, and bring to a boil. Reduce the heat to medium-low and cook until the potatoes are easily pierced with a fork or the tip of a sharp knife, 5 to 7 minutes. Drain and return to the pan.

Mash the potatoes slightly, then add the milk and finish mashing. You can leave them a bit lumpy if you like, or keep on going until they are smoother. Stir in the vegetables and beans and serve.

MAKES I SERVING

CALORIES: 540; STARCH: 3 (1 beans, 2 potatoes); VEGETABLES: 2½ (½ pepper, 2 spinach); FAT: 2 (olive oil); FRUIT: 0; DAIRY: ¼ (milk)

Vegetable-Stuffed Baked Potato

A plain baked potato becomes a satisfying, complete, and healthy meal when it is stuffed with vegetables. This easy dish is a favorite winter lunch when I am working from home.

1 medium (9-ounce) unpeeled baking potato, scrubbed clean
2 tablespoons extra virgin olive oil
¼ cup diced red onion
Salt and black pepper
½ cup sliced mushrooms
2 cups baby spinach or ⅓ cup thawed frozen chopped spinach

Preheat the oven to 400°F.

Pierce the potato in several places with a fork or skewer. Bake for about 1 hour, until it can be easily squeezed.

When the potato has been baking for about 40 minutes, start to cook the vegetables: Heat the olive oil in a medium skillet over medium-low heat. Add the onion, season with salt, and cook until soft, about 10 minutes. Stir in the mushrooms, season with salt, and cook until they begin to color, 5 to 7 minutes. Add the spinach, season with salt and pepper, and cook until it wilts; frozen spinach should cook for at least 7 minutes, or until any excess water has evaporated. The vegetables can continue to cook on low heat until the potato is done.

Slice the baked potato down the center and fill with the vegetables.

VARIATIONS: In addition to the mushrooms and spinach here, just about any vegetable—or combination of vegetables—cooked in olive oil makes a great topping for baked potatoes. Try broccoli with or without red peppers, shredded carrots and tender tiny peas, or chopped green cabbage and scallions. A couple of large cloves of garlic cooked with the onions in the recipe add a nice flavor, as does a tablespoon of chopped parsley added with the mushrooms.

MAKES I SERVING
CALORIES: 515; STARCH: 3 (potato); VEGETABLES: 3½ (½ onion, 1 mushrooms, 2 spinach); FAT: 2 (olive oil); FRUIT: 0; DAIRY: 0

Potato and Cauliflower Casserole with Cheese Sauce

This casserole makes a generous, satisfying meal. The cheese sauce is built in the pan with flour and milk. When you add the milk to the flour you have to stir quickly and thoroughly so that no flour remains and no lumps form. I like to use a whisk, but any tool that gets the job done is fine.

8 tablespoons extra virgin olive oil
2 cups chopped red onion (1 medium)
Salt
2 cups sliced raw cauliflower florets (1 small head)
1½ pounds unpeeled potatoes (Yukon Golds or russets), scrubbed clean
    and sliced thin (about ⅛ inch)
¼ cup all-purpose flour
1 cup nonfat or 1% milk, warmed
2 ounces shredded Cheddar cheese

Equipment needed: 13 × 9-inch shallow baking pan

Preheat the oven to 350°F.

Heat 6 tablespoons of the olive oil in a large pan over medium heat. Stir in the onion, season with salt and pepper, and cook just until translucent, about 10 minutes. Add the cauliflower, season with salt, and cook for 10 minutes, stirring occasionally so all the vegetables are mixed with the oil.

Meanwhile, prepare the potatoes: Cover the bottom of the baking pan with the potatoes, overlapping the slices slightly. Drizzle evenly with the remaining 2 tablespoons oil. Season with salt and pepper.

Sprinkle the flour over the vegetables and whisk in well so there is no dry flour visible. Whisk in the milk and continue whisking slowly until the mixture thickens slightly, 5 to 7 minutes. Add the cheese and cook until it melts.

Spread the vegetables and sauce evenly over the potatoes. Cover with foil and bake for 20 minutes. Remove the foil and bake for 20 to 25 minutes, until the potatoes are tender and can easily be pierced with the tip of a small, sharp knife or a fork. Use a knife to score the casserole into four even pieces.

MAKES 4 SERVINGS
CALORIES: 515; STARCH: 2 (potatoes; flour is too scant to be counted); VEGETABLES: 2 (1 each onion and cauliflower); FAT: 2 (olive oil); FRUIT: 0; DAIRY: ¾ (¼ milk, ½ cheese)

Potato, Pumpkin, and Cheese Casserole

Any cheese is fine for this recipe; smoked cheese is especially tasty. Be sure to use canned pumpkin that has no sugar or spices added.

1 tablespoon plus 2 teaspoons extra virgin olive oil
1 cup diced red onion (½ medium)
Salt and black pepper
1 tablespoon all-purpose flour
½ cup nonfat or 1% milk, warmed
¼ cup shredded Cheddar cheese (1 ounce)
½ cup canned pumpkin
1 small (6-ounce) unpeeled baking potato, scrubbed clean and thinly
    sliced (about ⅛ inch)

Equipment needed: small (6 × 8-inch) shallow baking dish

Preheat the oven to 350°F.

Heat 1 tablespoon of the olive oil in a small pan over medium heat. Stir in the onion, season with salt and pepper, and cook, stirring occasionally, until soft, about 5 minutes. Sprinkle the flour over the onion and whisk it in well so there is no dry flour visible. Whisk in the milk, and continue whisking until the mixture thickens slightly. Add the cheese and cook until it melts.

Spread ¼ cup of the pumpkin in the bottom of the baking dish. Top with a layer of half the potato slices. Season with salt, pepper, and 1 teaspoon of the olive oil. Spread slightly less than half of the cheese sauce over the potatoes. Repeat the layers using the remaining pumpkin, potatoes, olive oil, and cheese sauce.

Cover with foil and bake for 20 minutes. Remove the foil and bake for 20 to 25 minutes longer, until the top is lightly browned and you can easily pierce the potatoes with a fork.

MAKES I SERVING
CALORIES: 580; STARCH: 2 (potatoes; flour is too scant to be counted); VEGETABLES: 2 (I pumpkin, 2 onion); FAT: I⅔ (olive oil); FRUIT: 0; DAIRY: I½ (½ milk, I cheese)

Potato, Zucchini, and Fresh Tomato Casserole

This is particularly good in summer when the zucchini and tomato crops come in. It reheats nicely in the microwave, so if you want to increase the recipe, keep the extra servings refrigerated or frozen. Or have half the recipe below for lunch on two different days. It is quite satisfying accompanied by a small green salad. The olive oil used to coat the baking dish adds negligible calories, so it is not counted for fat.

2 tablespoons extra virgin olive oil, plus additional for greasing the
    baking dish
2 cups thinly sliced zucchini (2 medium)
Salt and black pepper
1½ cups sliced fresh tomatoes (about 8 ounces)
1 tablespoon all-purpose flour
½ cup nonfat or 1% milk, warmed

1 small (6-ounce) unpeeled baking potato, scrubbed clean and thinly
   sliced (about ⅛ inch thick)

Equipment needed: small (6 × 8-inch) shallow baking dish

Preheat the oven to 350°F. Use a pastry brush to coat the bottom and
sides of the baking dish with about ½ teaspoon olive oil.

Heat the 2 tablespoons olive oil in a medium pan over medium-low heat.
Stir in the zucchini, season with salt and pepper, and cook until it is
slightly soft, about 5 minutes. Add the tomatoes, season with salt and
pepper, and cook for 10 minutes. Don't worry if they fall apart. Sprinkle
on the flour and mix it in completely so there is no dry flour visible.
Whisk in the milk and continue whisking until the liquid starts to
thicken.

Arrange half of the potato slices in a single layer in the prepared dish.
The pieces can overlap slightly. Top with half the zucchini/tomato sauce.
Cover with the rest of the potatoes and spread on the remaining sauce.
Bake for 40 to 45 minutes, until the potatoes can be easily pierced with a
fork.

MAKES I SERVING
CALORIES: 540; STARCH: 2 (potato; flour is too scant to be counted); VEGETABLES: 7
(4 zucchini, 3 tomatoes); FAT: 2 (olive oil); FRUIT: 0; DAIRY: ½ (milk)

Beans

Beans (as well as lentils and peas) are legumes, that is, a species of
plants with seed pods that split along both sides when they are ripe—
and out pops the bean. (Peanuts are also legumes but nutritionally
closer to nuts, so we won't talk about them here). Legumes are an excellent source of protein, phytonutrients, essential micronutrients, and
fiber. They are rich in folic acid, copper, and magnesium—three nutrients few of us get enough of. If you are considering a vegetarian diet,
you should know that legumes are an extremely healthy alternative to
animal protein.

So why don't we eat more of these inexpensive, nutrient-dense little wonders? Beans are an essential food in the diets of many countries that enjoy lower incidences of chronic disease, but, oddly, not in our diets. Change that and start adding them to your meal plans, on a weekly if not daily basis. Add them anywhere—to soups, stews, salads, casseroles, sandwiches. Nancy saw a woman on *The Dr. Oz Show* who made brownies with black beans!

Most well-stocked grocery stores carry a good variety of packaged dried beans. Natural food stores usually sell them loose, in bulk. Already cooked beans, sold in cans, are in all markets. All varieties are allowed and I highly recommend that you try an array of types; there are so many compelling flavors and textures out there.

Dried beans will keep indefinitely in your cupboard (not the refrigerator, where the moisture could encourage mold). Keep them in the unopened bag and after opening, reclose the bag tightly with a twist tie or transfer them to a tightly closed glass or plastic container (along with the cooking directions torn from the bag to remind you of approximate cooking times). Preparing dried beans (page 128) requires you to plan ahead because they do need soaking time and then, depending on the variety, a fair amount of simmer time.

Okay, you want to know about that gassy situation beans can cause. All I'll say is that the more beans you consume, the more it will go away. Trust me.

Soaking and Cooking Dried Beans

Presoaking dried beans is not essential, but it does reduce cooking time. Some beans can take up to 3 hours to become tender; soaking, which requires no attention from you, can cut that time in half. (Lentils need no presoaking; they will cook in 30 to 35 minutes. See the recipe on page 149).

For any amount of beans, follow either of the two traditional presoaking methods given below, the Long Soak and the Quick Soak. Do not salt beans while they are soaking since it can toughen the skins, preventing them from becoming tender.

For either method, before soaking, spread the beans out on a large pan or in a big colander and go over them carefully to remove any small

---

GOOD TO KNOW

Bean Measurements

"A pint's a pound the world around" and a pint of dry beans (2 cups) indeed weighs a pound. After soaking and cooking, beans will have doubled or tripled in volume: 1 cup of dry beans swells to 2 or 3 cups cooked.

---

stones and badly shriveled, misshapen, hole-ridden, undersized, and discolored beans. Then rinse the beans well under running cold water, raking your fingers through them to remove all dirt. Drain.

1. *Long Soak:* Put the washed and drained beans in a large bowl or pot and pour in enough room-temperature water to come at least 2 inches above the beans. Cover the bowl and let sit (refrigerated if the kitchen is warm) for at least 4 hours and up to 12 hours (or overnight). Drain well and discard the soaking water.
2. *Quick Soak:* Put the washed and drained beans in a large pot or saucepan. Cover with water by 2 inches and bring to a boil. Boil for 2 minutes. Remove from the heat, cover the pot, and let the beans soak for 1 hour. Drain and discard the soaking water.

To Cook Beans: Put presoaked, drained beans in a large, heavy pot or saucepan and cover with fresh water by 2 inches. Bring to a boil, and then reduce the heat so the liquid barely simmers. (If you want, skim off any foam that rises to the top, but it is not harmful and eventually will be reabsorbed by the beans.) You can flavor the beans during cooking: For example, for about every 1 cup of beans, add a large celery stalk with its top, a medium onion, and a bay leaf; or a few whole cloves, some peppercorns, and perhaps parsley or thyme sprigs. Do not add salt until the beans start to soften or they never will. Simmer, covered with the pot lid at a tilt, until the beans are tender. You should be able to mash a cooked bean easily with your fingers. Beans can vary in cooking time from 30 minutes to 2 hours, depending on their variety and age. Check the beans occasionally as they cook and if they are not still covered with water, add very hot water to the pot.

Store cooked and cooled beans, in their liquid, in the refrigerator for up to 4 days. Freeze the cooked and cooled beans, in their liquid, in airtight containers. For the convenience of the diet, freeze in ½-cup allowance sizes and make sure there is room in the container for expansion of the beans and liquid. Cooked beans will keep in the freezer for about 3 months.

Black Bean, Onion, and Hot Pepper Dip/Spread

Don't be put off by the somewhat gray color of this dip. What it lacks in color, it makes up in flavor. Use as you would the Roasted Red Pepper Dip/Spread (page 131).

¼ cup extra virgin olive oil
¾ cup chopped red onion (slightly less than ½ medium)
¼ cup jarred hot pepper rings, minced
Salt and black pepper
1¾ cups cooked black beans (see page 128), or canned beans, drained
    and rinsed

Heat the olive oil in a medium pan over medium-low heat. Add the onion and chopped hot peppers, season very lightly with salt and pepper, and cook, stirring often, until very soft, 15 to 20 minutes. Stir in the black beans and cook until very soft, about 10 minutes. Cool to room temperature.

Transfer the beans and vegetables to a food processor fitted with a steel blade and process until smooth, 2 to 3 minutes.

MAKES 4 SERVINGS, ½ CUP EACH
CALORIES: 225; STARCH: slightly more than ¾ (beans); VEGETABLES: ½ (⅛ peppers, ⅜ onion); FAT: 1 (olive oil); FRUIT: 0; DAIRY: 0

Roasted Red Pepper Dip/Spread

This healthy version of red pepper hummus replaces the traditional (but not so healthy) sesame oil and tahini with olive oil. The flavor is still inviting: In fact, many people find this version more so. You'll see that I sauté the chickpeas before pureeing them. I did this once by mistake and found that it made the flavor of the dip much deeper. Having a party? The dip is a perfect accompaniment to raw vegetable crudités and Pita Chips (page 104). It is also good as a spread on sandwich wraps.

¼ cup extra virgin olive oil
1 garlic clove, minced
1½ cups roasted red bell peppers (see page 220) or 1 (7-ounce) jar red
    peppers, cut into strips
1 to 2 teaspoons ground cumin (optional)
Salt and black pepper
1¾ cups cooked chickpeas (see page 128), or canned chickpeas, drained
    and rinsed
2 tablespoons lemon juice

Heat the olive oil in a medium pan over medium heat. Add the garlic and cook until translucent but not brown, 2 to 3 minutes. Stir in the peppers, season with cumin (if using) and salt and pepper, and cook until most of the oil is absorbed, 8 to 10 minutes. Add the chickpeas and stir to coat. Continue cooking until the chickpeas start to brown slightly, 8 to 10 minutes. Let cool.

Transfer the cooled mixture to a food processer fitted with a steel blade. Add the lemon juice and process until smooth, about 3 minutes.

VARIATIONS: Add a few tablespoons of chopped flat-leaf parsley, or substitute hot red pepper flakes or chili powder for the cumin. I often use cannellini beans in place of the chickpeas because I like the sweeter taste and smoother texture.

MAKES 4 SERVINGS, 5 TABLESPOONS EACH
CALORIES: 220; STARCH: slightly more than ¾ (beans); VEGETABLES: ¾ (peppers);
FAT: 1 (olive oil); FRUIT: 0; DAIRY: 0

Cheesy Black Bean Nachos

There are a number of ways you can add these healthy nachos to your diet. One or two loaded triangles make a great appetizer or a nice side to a bowl of soup. You can enjoy four as lunch (double the calories and the serving allowances). Make all eight and serve them to company. If you don't make the entire recipe, toast only one tortilla and store the unused spread in the refrigerator for up to 3 days.

2 tablespoons extra virgin olive oil
½ cup diced red onion (½ small)
Salt
½ cup diced green bell pepper (½ medium)
½ cup favorite salsa
½ cup cooked black beans (see page 128), or canned beans, drained and
    rinsed
2 (6½-inch) whole wheat tortillas
½ cup shredded Cheddar cheese (2 ounces)

Heat the olive oil in a medium pan over medium-low heat. Stir in the onion, season with salt, and cook, stirring occasionally, until translucent but not brown, about 10 minutes. Add the pepper, season lightly with salt, and cook until starting to soften, about 5 minutes. The pepper should still be firm. Stir in the salsa and beans and cook for 5 minutes to blend the flavors.

Meanwhile, toast the tortillas: Preheat the broiler. Place the tortillas on a baking sheet (the pieces can touch but not overlap) and place under the broiler, turning once, until toasted. Remove from the oven—leave the broiler on—and cut each tortilla into four triangles. Return the tortillas to the baking sheet.

Divide the beans on top of the tortilla triangles, heaping them in mounds. Top each with 1 tablespoon of the Cheddar. Place under the broiler and cook until the cheese melts and is bubbly, 3 to 4 minutes.

MAKES 4 SERVINGS, 2 TRIANGLES EACH
CALORIES: 200; STARCH: ¾ (¼ beans, ½ tortilla); VEGETABLES: ¾ (¼ each onion, pepper, and salsa); FAT: ½ (olive oil); FRUIT: 0; DAIRY: ½ (cheese)

Baked Beans with Tomatoes

This makes a nice winter side dish as well as a great topping for pasta. You can also use the beans in the Bean, Spinach, and Roasted Red Pepper Sandwich (page 101). Plan ahead because the beans need to soak before cooking.

¼ cup extra virgin olive oil
2 cups chopped red onion (1 medium)
Salt and black pepper
1 (28-ounce) can diced or crushed tomatoes
6 or 7 fresh sage leaves
1 pound white navy beans, soaked 12 hours or overnight, or quick
    soaked (see page 128)

Equipment needed: Dutch oven or heavy pot with a lid

Preheat the oven to 250°F.

Heat the olive oil in the Dutch oven over medium heat. Stir in the onion, season with salt, and cook until beginning to soften, about 5 minutes. Stir in the tomatoes, season lightly with salt and pepper, and cook gently until warmed through. Stir in the sage leaves.

Add the soaked and drained beans to the tomatoes and pour in enough fresh water so the liquid is about an inch over the beans. Bring to a boil and boil for 7 minutes. Cover and transfer to the oven. Bake for 1 hour and stir. Continue to bake, covered, for 2½ to 3 hours, until the beans are tender and have absorbed all the liquid.

MAKES 16 SERVINGS, ABOUT ½ CUP EACH
CALORIES: 150; STARCH: 1 (beans); VEGETABLES: ¾ (¼ onion, ½ tomato); FAT: less than ¼;
FRUIT: 0; DAIRY: 0

Greek Bean and Vegetable Casserole

This is my version of a casserole I had on the last night of a trip to Greece. I was eating dinner alone on the island of Poros and ordered this. I was glad to be alone and not have to share it. It is good by itself, but can also be served over cooked brown rice or quinoa (see page 138). Remember to count the extra starch.

2 tablespoons extra virgin olive oil
¼ teaspoon dried red pepper flakes
¼ cup red onion
¼ cup diced celery
Salt and black pepper
1 cup cubed eggplant
1 cup sliced zucchini (1 medium)
½ cup diced red or green bell pepper (½ medium)
1 cup canned crushed tomatoes
1 bay leaf
½ cup cooked chickpeas (see page 128), or canned chickpeas, drained
   and rinsed
2 tablespoons feta cheese

Equipment needed: 6 × 8-inch casserole or baking dish

Preheat the oven to 375°F.

Heat the olive oil in a medium pan over medium heat. Add the dried red pepper, then the onion and celery. Season with salt and cook for 5 minutes. Add the eggplant, zucchini, and peppers and cook until some of the oil is absorbed into the vegetables, about 5 minutes. Add the tomato and bay leaf and cook until bubbling, 5 to 7 minutes. Stir in the chickpeas and transfer the mixture to the baking dish. Bake for 30 minutes. Sprinkle the cheese over the top. Return to the oven and bake for 5 minutes longer, or until the cheese melts. Discard the bay leaf before serving.

MAKES I SERVING
CALORIES: 540; STARCH: I (beans); VEGETABLES: 8 (½ each onion and celery; I peppers; 2 each eggplant, zucchini, and tomatoes); FAT: 2 (olive oil); FRUIT: 0; DAIRY: ½ (cheese)

Black Bean Veggie Burger

So easy to make and so good for you. Use it for the Veggie Burger Sandwich (page 103) or with Mashed Potatoes (page 226). If you want a "cheeseburger," put cheese on top to melt before removing from the pan. The recipe makes 6 veggie burgers; to freeze some for later use, tightly wrap each separately in plastic wrap, place in a freezer bag, and squeeze out excess air.

4 tablespoons extra virgin olive oil

1 cup diced red onion (½ medium)

1 tablespoon chili powder

Salt

2 cups diced red and/or green bell peppers, fresh or frozen (2 medium fresh)

2 cups cooked black beans (see page 128), or canned beans, drained and rinsed

1 large egg

1 cup cooked brown rice (see page 138)

¾ cup shredded Cheddar or mozzarella cheese (3 ounces)

Heat 2 tablespoons of the olive oil in a large skillet over medium-low heat. Stir in the onion, season with the chili powder and salt, and stir to distribute. Cook, stirring occasionally, until the onion is softened, about 7 minutes. Add the peppers and cook, stirring occasionally, until they are soft and most of the oil is absorbed by the vegetables, about 15 minutes. Let cool slightly.

While the vegetables are cooking, process the beans in a food processor fitted with a steel blade until partially smooth, 10 to 15 seconds; there should be some visible bean pieces. Add the egg and process until it is just mixed in, about 10 seconds.

Transfer the beans to a bowl. Gently mix in the cooled vegetables and the rice. With a ½-cup measure, scoop out 6 portions of the beans. Use your hands to gently work 2 tablespoons of the shredded cheese into each ½ cup. Form the portions into 6 patties, each about 4 inches in diameter and 1 inch high.

Heat the remaining 2 tablespoons olive oil in a large pan over medium heat. Add the patties and cook until browned on the bottom, about 5 minutes. Gently turn the patties over with a spatula and cook the other sides until browned, about 5 minutes longer. (To cook one patty, use 1 teaspoon oil and a small pan.) Let the cooked burgers set for about 2 minutes before serving.

MAKES 6 SERVINGS, 1 PATTY EACH
CALORIES: 275; STARCH: 1 (⅓ rice, ⅔ beans); VEGETABLES: 1 (⅓ onion, ⅔ peppers); FAT: ⅔ (olive oil); FRUIT: 0; DAIRY: ½ (cheese)

Grains

It wasn't all that long ago when recipes for bulgur or quinoa had to begin with instructions to first seek out your local health or natural foods store to buy the grains. Not so today. Whole grains have maneuvered their way into mainstream American food ways, parking themselves on supermarket shelves, right there next to their paler, highly processed cousins. Oh sure, there are some little-sought-after grains that are still only in specialty stores, but the most commonly cooked ones no longer require any detective work.

The grains we use in this book are brown rice, barley, bulgur, wheat berries, quinoa (pronounced *keen-wah*), and wild rice. (Wild rice is actually a grass but it is cooked as a grain; couscous, also usually treated like a grain, is actually pasta and a recipe for it is in the pasta section.) We have found all these whole grains in our local grocery store. If you have trouble finding them, try a natural food store or buy on the Internet; Bob's Red Mill (www.bobsredmill.com) is a good source that we have used.

Buying grains in bulk saves money, so check out large markets or natural food stores that have bulk bins. Sometimes the grains are fresher, sometimes not. It depends on how fast the turnover is. If a store sells lots of whole grains, most likely they are restocking often so their grains are likely to be fresher. Grains should look and smell new; older grains will look dull and very old ones will smell rancid.

To maintain freshness and discourage the possibility of infestations, store grains in tightly sealed containers in dry, dark locations away from the heat. If you purchase grains in bulk and take them home in a plastic bag, transfer them immediately to a container with a tight-fitting lid for storage; the plastic bag invites moisture and rancidity. If you buy pre-packaged grains, check the use-by date on the package and keep them in their wrappers until using, and then transfer unused grains to a tightly sealed container. It is convenient to put the cooking directions in the container with the grains so you don't have to check the book each time.

You can store grains in the refrigerator or freezer in warm weather, but then it is even more important that they be kept sealed to prevent the grain from absorbing moisture and odors and flavors from other foods stored in the refrigerator.

Because whole grains still have their outer layer, the germ, their shelf life is not as long as that of refined grains. The germ contains a small amount of fat which, like all fats, over time can become rancid. In general, properly stored whole grains will keep for about a year; brown rice about 6 months.

So you can always have a quick, nutritious meal at your fingertips, cook a quantity of grains on the weekend or when you have time. Cooked and cooled grains will keep in the refrigerator for up to 7 days when stored in a clean, covered container. When you take out a portion, wipe away any moisture that has collected on the cover. And be sure to use a perfectly clean utensil when you scoop out a portion so you don't introduce any bacteria that can lead to contamination and spoilage of the unused amount. (This also means that eating cooked grains out of the container and returning it to the fridge is a bad idea.)

For longer storage of cooked grains, you can freeze them for about 2 months. Put the cooled grains into a plastic bag, lay it flat on the counter, and squeeze out excess air; freeze the bag on its side. You can freeze in serving allowance sizes or put all the grain in one large bag; the frozen "slab" will break apart easily. If it doesn't, smack the bag on the counter a few times. We use frozen grains only for soups.

Cooking Whole Grains (Brown Rice, Pearl Barley, Wheat Berries, Wild Rice, Quinoa, Bulgur)

Basically, there are three methods for cooking grains: boiling, absorption, and steaming (we won't address steaming here). Boiling takes the guesswork out of the proportion of water to grain since you use much more water than you need (think cooking pasta). It also allows you more opportunity to taste the grain as it cooks to determine when it is done to your liking. In other words, it is more forgiving. Speaking of forgiveness, cooked or partially cooked grains will have none of it if you handle it with anything other than a fork; once it's fully cooked that means only fluffing, not stirring. Excess agitation releases the starch from the grain and leads to a mushy end product. Another *unforgiveable* omission is not draining boiled grains right away; leaving them to sit in their cooking water makes them soggy.

There is some trial and error to timing the cooking of grains, especially by the absorption method. Success depends on such variables as the degree of heat, the age and quality of the grain, and the altitude. Keep in mind that fiber-rich grains will maintain a pleasantly chewy texture after they are cooked—that's "tender" to a whole grain. Here are the estimated cooking times for any amount of popular presoaked grains:

Brown rice: 25 to 35 minutes
Pearl barley: 35 to 40 minutes
Wild rice: 40 to 55
Quinoa: 15 minutes
Bulgur: 12 to 15 minutes
Wheat berries: 1 hour and 10 minutes

Boiling Method: Rinse the grain in a sieve under cold running water to remove any dirt or debris. Presoaking is not strictly necessary, but if you are using a grain with a long cook time (over an hour, for example) you can reduce that time by as much as half by soaking the grain in cold water to cover for 8 hours or overnight. Drain, and proceed with the directions.

For 1 to 2 cups of grain, bring 3 to 4 quarts of water to a boil; add ½ to 1 teaspoon salt per cup of grain. Stir in the grain, return to a boil, then reduce the heat so the grain cooks at a very low boil, covered or not. Test the doneness at the low end of the cooking time indicated above. Drain in a strainer. If you are not eating the grain immediately, rinse in very cold water to stop its cooking.

Absorption Method: Rinse the grains, and presoak or not, as above. Pour two or three times as much water as grain into a saucepan or pot, using the package directions for suggestions. Bring the water to a boil. Add ½ to 1 teaspoon salt per cup of grain and use a fork to stir in the grain. Reduce the heat so the water is at a bare simmer, cover the pot, and cook undisturbed for the recommended time. Undisturbed means that you should not take the cover off and stir, or the water you so carefully proportioned to the grain will escape in the form of steam and you will wind up with a gummy grain. After the allotted cooking time, the water should be completely absorbed. If the grain is cooked but there is still some water, turn off the heat, cover the pan, and let it sit for 10 minutes; the grain will absorb the water.

Fiery Rice and Black Beans

Hot pepper rings give this dish a nice kick. Most supermarkets carry the rings (small, pickled chiles) in jars. They are sometimes called pepperoncini or Tuscan peppers.

2 tablespoons extra virgin olive oil
½ cup sliced red onion (½ small)
Salt
½ cup jarred hot pepper rings, cut in half
½ cup cooked black beans (see page 128), or canned beans, drained and
    rinsed
1 cup cooked brown rice or barley (see page 138)

Warm the olive oil in a medium pan over medium-low heat. Stir in the onion and pepper rings, season very lightly with salt and pepper, and cook until very soft, about 10 minutes. Add the beans, cook 2 to 3 minutes to heat, and then add the rice. Cook until the rice is warm, about 4 minutes longer.

MAKES I SERVING
CALORIES: 565; STARCH: 3 (I beans, 2 rice); VEGETABLES: 2 (I each onion and peppers);
FAT: 2 (olive oil); FRUIT: 0; DAIRY: 0

Rice and Spinach Frittata

*When Italians refer to one of these flat omelets, they almost always call it una bella frit-
tata, a beautiful frittata. They may call it bella because of the lovely, golden color or perhaps because of its versatility—it's good warm from the oven and at room temperature.
It loses none of its appeal even after two or three days in the refrigerator. And it's
portable: Italians consider it a picnic food. You can eat it as is or make a sandwich with it.
Any combination of vegetables works. It increases beautifully—so you can make several,
or use a large pan and make a larger frittata.*

2 tablespoons extra virgin olive oil
½ cup chopped red onion (½ small)
Salt and black pepper
2 cups baby spinach or ⅓ cup thawed frozen spinach
1 cup cooked brown rice (see page 138)
1 large egg
¼ cup nonfat or 1% milk

Equipment needed: small frying pan that can go from the stovetop to the
broiler (cast-iron is ideal)

Preheat the broiler.

Heat the olive oil in a small pan over medium heat. Stir in the onion,
season with salt and pepper, and cook until translucent, about 10 minutes. Add the spinach, season lightly with salt, and cook until wilted, 1
or 2 minutes; frozen spinach should be cooked for about 7 minutes, or
until any excess water has evaporated. Use a fork to gently stir in the
rice; keep warm.

Beat the egg in a small bowl with a fork. Add the milk and stir to thoroughly combine. Season with salt and pepper. Pour the egg and milk

over the vegetables. Lift the pan and tilt from side to side to distribute evenly. Cook over medium-low heat until the egg is set, about 5 minutes.

Put the pan under the broiler and broil until the top starts to brown, 3 to 4 minutes. Use a rubber spatula to loosen the frittata from the pan and slide onto a plate.

MAKES 1 SERVING
CALORIES: 555; STARCH: 2 (rice); VEGETABLES: 2 (1 onion, 2 spinach); FAT: 2 (olive oil); FRUIT: 0; DAIRY: 1¼ (¼ milk, 1 egg)

Rice with Corn, Black Beans, and Tomatoes

This is a colorful dish that kids really like. If you can't find canned Italian-style tomatoes, use any style chopped canned tomatoes and cook one minced garlic clove in the oil until golden before adding the corn; then add ½ teaspoon dried oregano and/or 2 tablespoons chopped fresh basil with the tomatoes.

2 tablespoons extra virgin olive oil
½ cup canned corn, drained, or thawed frozen corn
Salt
¼ cup canned black beans, rinsed and drained
1 cup canned Italian-style tomatoes
¾ cup cooked brown rice or barley (see page 138)

Heat the olive oil in a medium pan over medium-low heat and stir in the corn. Season lightly with salt and cook until the corn starts to brown, 8 to 10 minutes. Add the beans and cook for 5 minutes to flavor the beans with the oil. Add the tomatoes and cook for 5 minutes to blend the flavors. Stir in the rice and cook until warmed, about 4 minutes longer. Add the rice and cook until the rice is heated through, 4 to 5 minutes.

MAKES 1 SERVING
CALORIES: 550; STARCH: 2 (½ beans, 1½ rice); VEGETABLES: 3 (1 corn, 2 tomatoes); FAT: 2 (olive oil); FRUIT: 0; DAIRY: 0

Rice with Spinach and Cannellini Beans

This is a simple, quick dish that you can make with any vegetable in place of the spinach. If you want rice without the beans, deduct 110 calories and 1 starch count.

2 tablespoons extra virgin olive oil
½ cup chopped red onion (½ small)
½ cup chopped celery (½ medium stalk)
Salt
2 cups baby spinach or ⅓ cup thawed frozen spinach
½ cup cooked cannellini beans (see page 128), or canned beans,
    drained and rinsed
1 cup cooked brown rice (see page 138)

Heat the olive oil in a medium pan over medium-low heat. Stir in the onion and celery, season with salt, and cook until beginning to soften, about 8 minutes. Add the spinach, season with salt, and cook until it wilts, 1 or 2 minutes; frozen spinach should be cooked for 7 minutes, or until no excess water remains. Add the beans and cook for 5 minutes to blend flavors. Gently stir in the rice and cook until heated through, 4 to 5 minutes.

VARIATION: The dish is also good with tomatoes. Add 1 cup Italian-style tomatoes after the beans have cooked for 5 minutes. Let them simmer for 5 minutes before adding the rice. Add 2 vegetable servings to the count.

MAKES I SERVING
CALORIES: 575; STARCH: 3 (1 beans, 2 rice); VEGETABLES: 4 (1 each onion and celery, 2 spinach); FAT: 2 (olive oil); FRUIT: 0; DAIRY: 0

Vegetable Fried Rice with Egg

If you are an old hand at making fried rice with eggs, you probably have your own method. Some people cook the egg first until it is almost set, then scramble it and remove it from the pan or leave it in for the rest of the cooking time. Some add the egg at the very end, letting it scramble into the rice. I like my way below, but you can use any method with which you are comfortable. There is one point on which all cooks agree: The only rice to use is leftover, cooked, and cooled rice. Speaking of leftovers, fried rice is a great way to use up leftover vegetables.

2 tablespoons extra virgin olive oil
½ cup thinly sliced or shredded carrot (1 medium)
1 cup snow peas
½ cup sliced mushrooms
Salt and black pepper
1 large egg, beaten
1 cup cooked brown rice (see page 138)

Equipment needed: A wok feels right but is not necessary; a skillet will do, but it should be nonstick.

Heat 1 tablespoon of the olive oil in a small wok or medium nonstick skillet over medium heat. Add the carrot, pea pods, and mushrooms and season with salt and pepper. Cook until the vegetables are as tender as you like, 10 to 12 minutes.

Push the vegetables well to the side of the pan and add the egg. Wait for it to begin to set, 2 to 3 minutes, then use a fork to break it up and mix with the vegetables. Push the vegetables aside again. (Or, if you have increased the recipe and the vegetables take up too much room to be off to the side, remove them from the pan and return after the rice is fried.)

Turn the heat to medium-high and add the remaining 1 tablespoon oil to the pan's empty spot. Add the rice and stir into the oil. Continue to "fry" until the rice is heated through, about 3 minutes. Gently mix the rice and vegetables together and serve.

VARIATION: Cook a quarter-size slice of fresh ginger, with or without 1 sliced garlic clove, maybe some chopped scallions, in the oil before adding the carrots. Pour in 1 tablespoon reduced-sodium soy sauce just before serving. Top with more chopped scallions.

MAKES I SERVING
CALORIES: 595; STARCH: 2 (rice); VEGETABLES: 4 (I each carrots and mushrooms, 2 pea pods); FAT: 2 (olive oil); FRUIT: 0; DAIRY: I (egg)

Rice-Stuffed Bell Pepper

Use a green, red, yellow, or orange pepper. If you are increasing the recipe, choose different color peppers for a cheery appearance. The pepper can be baked well in advance, even a day or two ahead of time. It is good warm or at room temperature. You can also assemble the pepper ahead of time and bake just before you are ready to eat.

1 medium bell pepper, stem removed
Salt
2 tablespoons extra virgin olive oil
¼ cup diced red onion
¼ cup fresh flat-leaf parsley leaves
½ cup cooked brown rice (see page 138)

Equipment needed: small baking dish

Preheat the oven to 350°F. Bring a large saucepan of water to a boil.

Slice the top off the pepper just below the curve. Dice the top piece and set aside. Slice the pepper in half, from stem end to the bottom, so you have 2 sides that can be filled. Remove the seeds and ribs. When the water is boiling, add a pinch of salt and then drop in the pepper halves. Cook for 3 minutes. Remove with a slotted spoon or drain in a sieve.

Put the pepper halves in the baking dish and brush with a small amount of the olive oil; use about ½ teaspoon total. Turn the peppers so they are cut sides up.

Heat the remaining olive oil in a small pan over medium-low heat. Stir in the onion, season with salt, and cook until beginning to soften, 3 to 5 minutes. Add the diced pepper and cook until beginning to soften, 5 to 7 minutes. Stir in the parsley, cook for a minute to release its flavor, then gently stir in the rice.

Divide the mixture between the 2 pepper halves. Cover loosely with foil and bake for 20 minutes. Remove the foil and turn on the broiler. Broil the peppers until the pepper edges are slightly browned.

MAKES I SERVING

CALORIES: 380; STARCH: 1 (rice); VEGETABLES: 3 (½ each onion and parsley, 2 pepper); FAT: 2 (olive oil); FRUIT: 0; DAIRY: 0

Rice with Roasted Zucchini

This is a nice, light summer meal. I've tried it with dried rosemary and no matter how little I used the flavor was too strong. So if you don't have fresh rosemary—unless you are incredibly passionate about this dried herb—substitute another herb.

2 cups sliced zucchini (2 small; about 8 ounces total)
3 or 4 sprigs fresh rosemary
Salt and black pepper
2 tablespoons extra virgin olive oil
1 cup cooked brown rice (see page 138)
2 tablespoons grated Parmesan cheese

Equipment needed: baking dish large enough to hold the zucchini in one layer

Preheat the oven to 425°F.

Place the zucchini in the baking dish, drop in the rosemary, and season with salt and pepper. Add the olive oil and use a rubber spatula to turn the slices over and around until the zucchini is well-coated. Spread the zucchini out into one layer. Roast until the zucchini is tender and lightly browned, about 40 minutes. Halfway through roasting, use a fork to turn each slice over. (It is best to remove the pan from the oven and close the oven door so the oven remains at a high temperature.) Discard the rosemary sprigs.

Add the rice to the zucchini, stirring gently so the rice gets coated with the oil in the bottom of the dish. Sprinkle with the cheese and serve.

MAKES 1 SERVING
CALORIES: 530; STARCH: 2 (rice); VEGETABLES: 4 (zucchini); FAT: 2 (olive oil); FRUIT: 0; DAIRY: ½ (cheese)

Barley with Zucchini and Fresh Tomatoes

The texture of shredded zucchini after it has cooked makes it seem like an entirely different vegetable than sliced zucchini. It also cooks very quickly, so this is a good recipe to have at hand. I've used it here with barley, but it is also good with cooked brown rice or bulgur. And try it tossed with cooked whole wheat pasta.

2 tablespoons extra virgin olive oil
1 cup shredded zucchini (1 small; about 9 ounces)
Salt
2 cups chopped fresh tomatoes (9 to 10 ounces)
¼ cup fresh basil leaves, torn into small pieces
1 cup cooked barley (see page 138)

Heat the olive oil in a medium pan over medium-low heat. Stir in the zucchini, season with salt, and cook until soft, 3 to 5 minutes. Add the tomatoes, season with salt, and add the basil. Cover the pan and cook until the tomatoes break up, 3 to 5 minutes. Stir in the barley, heat through, and serve.

VARIATIONS: Cook a finely chopped fat garlic clove in the oil until translucent before adding the zucchini. Substitute fresh mint leaves for the basil.

MAKES I SERVING
CALORIES: 545; STARCH: 2 (barley); VEGETABLES: 6 (2 zucchini, 4 tomatoes);
FAT: 2 (olive oil); FRUIT: 0; DAIRY: 0

Quinoa with Carrots and Peas

Quinoa is an ancient grain, a staple to the Incas who called it "the mother grain." This delicately flavored grain is a complete protein; that is, it contains all eight essential amino acids. Ivory-colored quinoa is the most available variety, but many natural foods stores carry the black and red types. They are all cooked in the same way, like rice: either by stirring into boiling water or, for a more defined texture, by first sautéing in oil and then adding water and boiling (pilaf method). Quinoa is fun and "in," so play with it. Sub-

---

GOOD TO KNOW

Barley

Perhaps barley, a grain that dates back to the Stone Age, owes its longevity of appeal to its pleasant, roasted-nut flavor. The most nutritious form of barley is whole grain (hulled or Scotch) barley, which has only the outer husk removed. What is commonly sold in markets is the off-white, oval variety known as pearl barley that has had the husk, bran, and germ removed, leaving the endosperm. Because the tougher components have been stripped away, pearl barley needs no presoaking, cooks much faster, and becomes more tender than whole barley. Primarily for those reasons, we tested the recipes in this book with pearl barley. Note that it is a good substitute in recipes that call for brown rice.

---

stitute different vegetables in the recipe below, use different herbs and seasonings, have a ball.

1 tablespoon extra virgin olive oil
½ chopped red onion (½ small)
Salt
½ cup thinly sliced carrot (1 medium)
¾ cup low-sodium vegetable broth
¼ cup plus 2 tablespoons quinoa
2 bay leaves
¼ cup thawed frozen peas

Heat the olive oil in a 1-quart pan over medium-low heat. Add the onion, season with salt, and cook until soft, 7 to 8 minutes. Add the carrot and cook until starting to soften, 8 to 10 minutes.

Add the vegetable broth, increase the heat to medium-high, and bring to a boil. Add the quinoa and bay leaves, cover, and reduce the heat to low. Simmer for 20 minutes, or until all the liquid is absorbed. Discard the bay leaves. Stir in the peas and serve.

MAKES 1 SERVING
CALORIES: 430; STARCH: 2 (quinoa); VEGETABLES: 2½ (½ peas, 1 each onion and carrot); FAT: 1 (olive oil); FRUIT: 0; DAIRY: 0

Bulgur with Roasted Green Beans and Tomatoes

Canned Italian-style tomatoes are good here; if you use them you won't need the dried oregano. If you have a large crop of your own green beans, increase the recipe to accommodate your haul. The dish freezes well and will keep in the refrigerator for several days.

2 tablespoons extra virgin olive oil
1 cup fresh green beans cut in half (about 4 ounces) or
    thawed frozen green beans
1 cup diced or crushed canned tomatoes with liquid
½ to 1 teaspoon dried oregano or ¼ teaspoon ground cloves
Salt
2 ounces bulgur wheat

Equipment needed: small roasting pan

Preheat the oven to 425°F.

Toss the olive oil, green beans, and tomatoes together in a bowl. Season with oregano and salt. Transfer to the roasting pan and roast for 30 to 40 minutes, until most of the tomato liquid is gone and the beans are browned.

Meanwhile, bring 1 cup of water to a boil, add a pinch of salt, and stir in the bulgur. Cover the pan and reduce the heat. Simmer for 12 to 15 minutes, until the bulgur is tender. Drain off any excess water.

Mix the bulgur into the green beans and serve.

VARIATION: In place of the bulgur, use 1 cup cooked brown rice, adding it as soon as the green beans come out of the oven. The heat of the dish will warm the rice.

MAKES I SERVING
CALORIES: 530; STARCH: 2 (bulgur); VEGETABLES: 4 (2 each green beans and tomatoes);
FAT: 2 (olive oil); FRUIT: 0; DAIRY: 0

Lentils

Have these with Mashed Potatoes (page 226).

1 tablespoon extra virgin olive oil
½ cup diced red onion (½ small)
Salt
½ cup very thinly sliced carrot (1 medium)
¼ cup brown lentils
2 or 3 bay leaves

Heat the olive oil in a medium pan over medium-low heat. Add the onion, season with salt, and cook until translucent but not brown, about 10 minutes. Stir in the carrot and cook until soft, 8 to 10 minutes.

Raise the heat to medium-high and add the lentils, bay leaves, and 1 cup hot water. Cover and bring to a boil. Reduce the heat to medium-low and cook until all the water is absorbed, 30 to 40 minutes. Discard the bay leaves before serving.

SERVES I
CALORIES: 330; STARCH: 2 (lentils); VEGETABLES: 2 (I each onion and carrots);
FAT: I (olive oil); FRUIT: 0; DAIRY: 0

Pasta

Pasta, like beans, grains, and potatoes, is a starch, but it gets a chapter all to itself—as well it should. Fiber-rich whole wheat pasta is the ideal food for a Mediterranean-style diet. It's a worthy companion to all vegetables and has a robust flavor that marries well with even the most full-bodied olive oil. What's not to like?

Well, that is kind of the problem. Many people, including women in my program and even Nancy, said that they just couldn't eat whole wheat pasta. I understood because it wasn't all that many years ago that I couldn't either. I tried all the brands available to me and decided that pasta plus whole wheat equaled gummy, gritty-tasting gobs that refused to hold a sauce.

The problem was simple. Those pastas were not being made by the people who—forget about Marco Polo and China—invented it, the Italians. Good Italian cooks will tell you that there is no substitute for pasta made in the Italian tradition—using Italian, hard durum wheat, shaped in bronze dyes, and allowed to dry slowly. The same is true for whole wheat pasta. Durum wheat allows the pasta to boil without becoming gummy; bronze dyes, unlike the more commonly used Teflon dyes, give the pasta a textured surface that holds a sauce; and slow drying at low temperatures preserves the nutty flavor of the wheat.

This pasta story has a happy ending. Good, Italian whole wheat pasta is now available, if not in your local market or natural food store, then on the Internet. Two that we really like are Bionaturae (www.bionaturae .com) and the slightly bolder-flavored DeLallo (www.delallo.com). Both companies produce a variety of shapes; DeLallo makes whole wheat, no-boil lasagna noodles.

Cooking whole wheat pasta follows the same rules as all pasta: use plenty of water and add salt to the water until the water tastes pleasantly salty. (But adding oil to the boiling water is unnecessary and can actually hurt by making the pasta too slippery to hold the sauce.) Stir in the pasta and keep it moving until the water returns to a boil, cook uncovered, stirring occasionally, drain, and sauce right away. Before draining the pasta, it is prudent to remove about a cup of the cooking water in case your sauce needs a little "stretching." The pasta water has residual starch that will help bind the sauce.

There are plenty of recipes in this section for pasta sauces and dishes, but I hope you'll make up your own with what you have in the garden or vegetable bin. You'll quickly see how easy and fast it is to make a healthy meal. If you have leftovers from another meal, more than likely they will be great on pasta.

The pasta recipes were popular with the women in my study especially for serving to families and company. All the recipes can be increased by multiplying the ingredients by the number of people; you may want to increase it even more for bigger eaters. Most of the pasta dishes begin with 2 tablespoons of olive oil; if you want to have a salad on the side you can cut it down to 1 tablespoon. If the pasta is all you are eating for dinner, do keep it at 2 so you will not be hungry later.

Tomato Sauce

This is an all purpose tomato sauce—cheaper to make than what you'll find on the grocery store shelf and with nothing in it but what needs to be there. Freeze the sauce in ½-cup portions (enough for a 2- or 3-ounce serving of pasta) and you'll always have a "fast food" meal on hand.

¼ cup extra virgin olive oil
1 cup finely chopped yellow onion (1 medium)
3 large garlic cloves, finely chopped
½ cup chopped fresh flat-leaf parsley
1 to 1½ teaspoons dried oregano or 1 to 1½ tablespoons chopped fresh
   oregano
¼ teaspoon sweet paprika
Salt and black pepper

½ cup red wine
7 cups canned crushed tomatoes (two 28-ounce cans)

Heat the olive oil in a large deep pot over medium-low heat. Add the onion, garlic, parsley, and oregano, stir, then season with paprika and a little salt and pepper. Cook slowly until the vegetables are tender, about 10 minutes; do not brown. Turn the heat up to high, immediately pour in the wine, and boil until it has reduced to a few tablespoons. Stir in the tomatoes and season with a little more salt and pepper. Bring to a boil and then reduce the heat to medium-low. Cover the pan with the lid slightly askew so the sauce does not reduce too quickly and simmer for 30 minutes.

The sauce will keep in the refrigerator for up to 4 days. Freeze in ½- or 1-cup portions for 4 months.

VARIATIONS: For a spicier sauce, cook ½ teaspoon hot pepper flakes with the onions; use hot not sweet paprika and use up to 1 teaspoon. Paprika is in the less gutsy sauce because it perks up the flavor of the tomatoes, not for the heat.

MAKES ABOUT 11 SERVINGS, ABOUT ½ CUP EACH
CALORIES: 105; STARCH: 0; FAT: about ⅓ (olive oil); VEGETABLES: about 1½
(less than ½ parsley, about 1/5 onion, about 1 tomatoes); FRUIT: 0; DAIRY: 0

Basil Pesto with Walnuts

Traditional basil pesto uses pignoli nuts, not walnuts, but I think you'll like the flavor and the extra nutrients that walnuts add. Like any pesto (see Arugula Pesto, page 154), this freezes nicely, but you are best off not adding the cheese until you thaw the sauce. Dry grating cheeses such as Parmesan and Asiago don't like to be that cold. I freeze pesto in muffin liners in serving size portions. To use it on pasta, thaw a serving size, add a tablespoon of cheese, and put it in a skillet large enough to hold the cooked pasta. Just before draining the pasta, add a scant ¼ cooking water to the pesto. Stir it around a bit to smooth it out before tossing in the pasta. If you forgot to add the cheese back to the pesto, sprinkle it on top of the pasta. Note that pesto is also good as a dip for vegetables or Pita Chips (page 104).

¼ cup extra virgin olive oil
1 garlic clove, minced (by hand or through a garlic press)
2 cups fresh basil leaves, rinsed and drained
¼ cup walnuts
¼ cup grated Parmesan or Asiago (1 ounce)
¼ teaspoon salt

Heat the olive oil with the garlic in a small pan over low heat until the garlic is just golden, about 3 minutes. Remove from heat and cool slightly.

Combine the garlic, basil, walnuts, Parmesan, and salt in a food processor fitted with a steel blade. Process for 30 seconds. Scrape down the sides and process 1 to 2 minutes longer, until the ingredients are blended.

MAKES 4 SERVINGS, ¼ CUP EACH
CALORIES: 190; STARCH: 0; VEGETABLES: ½ (basil); FAT: 1½ (½ walnuts, 1 olive oil); FRUIT: 0; DAIRY: ¼ (cheese)

Arugula Pesto

Arugula pesto is a nice alternative to Basil Pesto (page 153) when basil is out of season. The techniques and uses for it are the same. For a parsley version, substitute 4 cups loosely packed, flat-leaf parsley leaves for the arugula and ¼ cup walnuts for the almonds. The calories and serving sizes remain the same.

¼ cup extra virgin olive oil
1 garlic clove, minced (by hand or through a garlic press)
4 cups arugula, rinsed, drained and dried
¼ cup slivered almonds
¼ cup grated Parmesan or Asiago cheese (1 ounce)
1 tablespoon lemon juice
¼ teaspoon salt

Heat the olive oil with the garlic in a small pan over low heat until the garlic is just golden, about 3 minutes. Remove from heat and cool slightly.

Combine the garlic, arugula, almonds, cheese, lemon juice, and salt in a food processor fitted with a steel blade. Process for 30 seconds. Scrape down the sides and process 1 to 2 minutes longer, or until the ingredients are well blended.

MAKES 4 SERVINGS, ¼ CUP EACH
CALORIES: 200; STARCH: 0; VEGETABLES: 1 (arugula); FAT: 1½ (½ almonds, 1 olive oil); FRUIT: less than ¼ (lemon juice) ; DAIRY: ¼ (cheese)

Pasta with Spinach and Gorgonzola

I prefer to use fresh spinach for this recipe because it looks better, but if you need a quick meal and you only have frozen, go ahead and use it. Who else will know? You'll need ⅓ cup, thawed. The sauce goes together so quickly that I start the pasta cooking first.

2 ounces whole wheat pasta
Salt and black pepper
2 tablespoons extra virgin olive oil
2 cups baby spinach
1 ounce Gorgonzola cheese

Bring 6 cups water to a boil. Add salt and stir in the pasta. Continue to stir slowly until the water returns to a boil so the pasta does not stick together. Cook according to the time listed on the pasta package, tasting a minute before in case the package is wrong.

Meanwhile, heat the olive oil in a medium skillet over medium-low heat. Add the spinach, season with salt and pepper, and cook until wilted, 1 or 2 minutes. Add the Gorgonzola and 2 tablespoons of the pasta cooking water and stir until the cheese melts. Remove from the heat.

Drain the pasta and toss with the spinach and cheese.

MAKES 1 SERVING
CALORIES: 525; STARCH: 2 (pasta); VEGETABLES: 2 (spinach); FAT: 2 (olive oil); FRUIT: 0;
DAIRY: 1 (cheese)

Pasta with Spinach, Feta, and Olives

The feta and the olives are salty, so take care with how much salt you add to the dish.

2 tablespoons extra virgin olive oil
1 garlic clove, minced (by hand or through a garlic press)
2 cups baby spinach or ⅓ cup thawed frozen spinach
Salt
10 small or 6 large pitted black olives, cut in half (about 1 ounce)
½ ounce (2 tablespoons) feta cheese
2 ounces whole wheat pasta

Heat the olive oil in a medium skillet over medium-low heat. Stir in the garlic and cook just until golden but not browned, about 5 minutes. Stir in the spinach, season with salt, and cook until wilted, 1 or 2 minutes; frozen spinach should be cooked for 7 minutes, or until any excess water has evaporated. Stir in the olives and feta and remove from the heat.

Meanwhile, bring 6 cups water to a boil. Add salt and stir in the pasta. Continue to stir slowly until the water returns to a boil so the pasta does not stick together. Cook according to the time listed on the pasta package, tasting a minute before in case the package is wrong. Before draining, determine if the sauce needs more liquid and add up to ¼ cup pasta cooking water to it.

Drain the pasta and toss with the sauce.

MAKES 1 SERVING
CALORIES: 520; STARCH: 2 (pasta); VEGETABLES: 2 (spinach); FAT: 2½ (½ olives, 2 olive oil);
FRUIT: 0; DAIRY: ½ (cheese)

Spaghetti with Tomatoes, Feta, and Basil

Out of tomato season, I find that cherry or grape tomatoes are a good substitute for garden fresh tomatoes because they are usually sweet and juicy. Fresh basil is available year-round in most grocery stores, but oregano is a good combination with the tomatoes and feta. Use about 1 tablespoon fresh or 1 teaspoon dried oregano leaves and add them to the tomatoes as they cook.

2 tablespoons extra virgin olive oil
1 cup cherry or grape tomatoes, cut in half (about 5 ounces)
Salt
1 ounce feta cheese
¼ cup torn fresh basil leaves
2 ounces whole wheat spaghetti

Heat the olive oil in a medium skillet over medium-low heat. Add the tomatoes, season with salt, and cook slowly so they absorb the olive oil, about 10 minutes. Add the feta and cook until it starts to melt, 2 to 3 minutes. Remove from the heat and add the basil.

Meanwhile, bring 6 cups water to a boil. Add salt and stir in the pasta. Continue to stir slowly until the water returns to a boil so the pasta does not stick together. Cook according to the time listed on the pasta package, tasting a minute before in case the package is wrong.

Drain the pasta and toss with the sauce.

MAKES 1 SERVING
CALORIES: 520; STARCH: 2 (spaghetti); VEGETABLES: 2 (tomatoes); FAT: 2 (olive oil); FRUIT: 0; DAIRY: 1 (cheese)

Pasta with Fresh Tomato Sauce

"Simplicity is the ultimate sophistication." So said Leonardo da Vinci and—Mona Lisa be damned—my guess is he was sitting down to a plate of spaghetti with fresh tomato and basil sauce when he said it. If so, the tomatoes and basil that dressed his pasta had been picked moments before they went into the pan. This recipe is at its very best when the tomatoes and basil have been sun-ripened in your own garden or by a local grower. Fresh basil is available year-round in most grocery stores, but if you have a sunny spot in your garden, put in a few basil plants. They are easy to grow and will reward you abundantly.

2 tablespoons extra virgin olive oil
1 large garlic clove, peeled and thinly sliced (optional)
2 cups chopped fresh tomatoes (9 to 10 ounces)
Salt
¼ cup torn fresh basil leaves
2 ounces whole wheat pasta (any shape)
2 tablespoons grated Parmesan cheese

Heat the oil and garlic in a saucepan over medium-low heat until the garlic is golden but not brown, about 8 minutes. For a subtle garlic flavor, discard the garlic at this point. Add the tomatoes, season with salt, and cook gently, stirring occasionally, until they release their juices, 10 to 15 minutes. Stir in the basil and keep the sauce warm.

Meanwhile, bring 6 cups water to a boil. Add salt and stir in the pasta. Continue to stir slowly until the water returns to a boil so the pasta does not stick together. Cook according to the time listed on the pasta package, tasting a minute before in case the package is wrong.

Drain the pasta and transfer to a pasta bowl. Spoon the tomatoes on top and sprinkle with the cheese. *Buon appetito!*

MAKES 1 SERVING
CALORIES: 515; STARCH: 2 (pasta); FAT: 2 (olive oil); VEGETABLES: 4½ (½ basil and garlic, 4 tomatoes); FRUIT: 0; DAIRY: ½ (cheese)

Linguine with Green Beans and Tomato Sauce

I created this recipe, and others that use green beans with pasta, the summer I had a bumper crop of beans. You can substitute frozen green beans if fresh are out of season.

2 tablespoons extra virgin olive oil
½ cup diced red onion (½ small)
Salt
1 cup fresh green beans cut into 2-inch pieces (about 4 ounces)
½ cup Tomato Sauce (page 152) or your favorite meatless jarred sauce
2 ounces whole wheat linguine (or any whole wheat pasta of your choice)

Heat the olive oil in a medium pan over medium-low heat. Stir in the onion, season with salt, and cook, stirring occasionally, until softened, about 7 minutes. Add the green beans and cook until the beans can be easily pierced with a fork, about 15 minutes. Add the tomato sauce, heat through, and keep warm.

Meanwhile, bring 6 cups water to a boil. Add salt and stir in the pasta. Continue to stir slowly until the water returns to a boil so the pasta does not stick together. Cook according to the time listed on the pasta package, tasting a minute before in case the package is wrong.

Drain the pasta, transfer to the pan with the sauce, and toss together.

VARIATIONS: This recipe is the way I begin all quick vegetable and tomato pan sauces, and you can vary it in hundreds of ways: cook herbs, spices, and/or one or two coarsely chopped garlic cloves with the onions. In place of the green beans use 1 cup sliced peppers (fresh or frozen; green, red, or a combination) or ½ cup each of mushrooms and fresh or frozen peas—or a whole cup of either. You can also substitute 1 cup of fresh chopped tomato for the tomato sauce (count as an additional serving of vegetables).

MAKES I SERVING
CALORIES: 535; STARCH: 2 (pasta); VEGETABLES: 4 (I each onion and tomato sauce, 2 green beans); FAT: 2 (olive oil); FRUIT: 0; DAIRY: 0

Pasta with Green Beans and Lemony Ricotta

This is a nice company dish—just increase the ingredients by the number of people to serve. It looks pretty and tastes great. Skip the red pepper if you like; the pasta will still look pretty and taste great. You can begin the dish with fresh red bell peppers—they will lend a different, but still good, flavor than roasted peppers do—just be sure to cook them in the oil longer so they soften and absorb the flavor of the oil. Pay attention to the directions for cooking the pasta because you need some of the water before you drain it.

¼ cup part-skim ricotta cheese
1 tablespoon Parmesan cheese
1 teaspoon grated lemon rind (optional)
2 teaspoons lemon juice
1 tablespoon extra virgin olive oil
1 cup fresh green beans cut into 2-inch pieces (about 4 ounces)
Salt
½ cup sliced roasted red bell pepper (see page 220) or jarred red pepper
2 ounces whole wheat pasta (any shape)

Mix both cheeses together and then mix in the lemon rind (if using) and juice. Beat with a fork until the mixture is smooth. Set aside.

Heat the olive oil in a medium skillet over medium heat. Stir in the green beans, season with salt, and cook until tender, about 15 minutes. Add the pepper, stir around to coat with oil, and cook until heated through, about 5 minutes. Remove the pan from the heat and stir in the lemon ricotta.

Meanwhile, bring 6 cups water to a boil. Add salt and stir in the pasta. Continue to stir slowly until the water returns to a boil so the pasta does not stick together. Cook according to package instructions, tasting a minute before in case the package is wrong. When the pasta is a minute or two away from being cooked, remove ¼ cup of the cooking water and stir it into the cheese and beans to stretch and smooth out the sauce.

Drain the pasta and toss it with the sauce.

MAKES I SERVING
CALORIES: 460; STARCH: 2 (pasta); VEGETABLES: 3 (I pepper, 2 green beans);
FAT: I (olive oil); FRUIT: 0; DAIRY: 1¼ (¼ Parmesan, I ricotta)

Linguine with Green Beans and Walnuts

Walnuts add a satisfying crunch to this pasta. Be sure and add the lemon juice just before tossing in the pasta or the acid could rob the beans of their color. You can substitute frozen green beans.

2 tablespoons extra virgin olive oil
1 cup fresh green beans cut in half (about 4 ounces)
Salt
1 tablespoon chopped walnuts
2 ounces whole wheat linguine
1 tablespoon lemon juice

Heat the olive oil in a medium skillet over medium heat. Add the beans, season with salt, and cook, turning occasionally, until browned, about 15 minutes. Stir in the nuts and cook a minute or two, until they absorb the olive oil flavor. Remove the pan from the heat if you are not ready to finish the dish.

Meanwhile, bring 6 cups water to a boil. Add salt and stir in the pasta. Continue to stir slowly until the water returns to a boil so the pasta does not stick together. Cook according to the time listed on the pasta package, tasting a minute before in case the package is wrong.

Add the lemon juice to the beans and nuts. Drain the pasta and toss with the beans and nuts.

MAKES I SERVING
CALORIES: 500; STARCH: 2 (pasta); VEGETABLES: 2 (green beans); FAT: 2½ (½ walnuts,
2 olive oil); FRUIT: 0; DAIRY: 0

Pasta with Spicy Spinach-Caper Sauce

As I have written before, when a dish has a lot of flavor you are more satisfied than you would be with a bland meal, and consequently less likely to overeat. This recipe has plenty of taste, from the heartiness of the whole wheat pasta to the spice and capers. The sauce is low in calories and enough for 3 ounces of pasta, so if your daily intake can accommodate another starch, increase the pasta to 3 ounces and add 100 calories and 1 starch.

2 tablespoons extra virgin olive oil
½ cup sliced red onion (½ small)
1 garlic clove, minced (by hand or through a garlic press)
¼ to ½ teaspoon crushed red pepper flakes
2 cups baby spinach or ⅓ cup thawed frozen spinach
Salt
½ cup canned diced tomatoes with liquid
1 tablespoon capers, rinsed
2 ounces whole wheat pasta

Heat the olive oil in a medium skillet over medium-low heat. Stir in the onion, garlic, and red pepper flakes and cook just until the garlic is golden and the onion soft, about 7 minutes; do not let them brown. Add the spinach, season lightly with salt, and cook until wilted, 1 or 2 minutes; frozen spinach should be cooked for 7 minutes, or until any excess water has evaporated. Stir in the tomatoes and capers, let the sauce bubble for a minute, then reduce the heat and keep warm.

Meanwhile, bring 6 cups water to a boil. Add salt and stir in the pasta. Continue to stir slowly until the water returns to a boil so the pasta does not stick together. Cook according to the time listed on the pasta package, tasting a minute before in case the package is wrong.

Drain the pasta and toss with the sauce.

SERVES 1
CALORIES: 465; STARCH: 2 (pasta); VEGETABLES: 4 (1 each onion and tomato, 2 spinach); FAT: 2 (olive oil); FRUIT: 0; DAIRY: 0

Spicy Pasta and Beans

Beans and pasta have had a long and successful marriage in many cultures. It's not surprising; together they supply a blissful amount of fiber and nutrients, not to mention satiety. The hot peppers add a nice zing to the sauce—to any sauce for that matter. Add more if you like, or if the zing thing flings you for a loop, omit them altogether. A little grated Parmesan or Romano is nice on top—if you add it, count the dairy.

2 tablespoons extra virgin olive oil
¼ cup jarred hot pepper rings
1 cup chopped fresh tomatoes (9 to 10 ounces)
Salt
½ cup cooked cannellini beans (see page 128), or canned beans, drained
    and rinsed
2 ounces whole wheat pasta (any shape)
¼ cup fresh basil leaves, torn into small pieces

Heat the olive oil in a medium pan over medium heat. Add the hot pepper rings and cook until lightly browned, about 5 minutes. Stir in the tomatoes, season with salt, and reduce the heat to medium-low. Cook until most of the oil is absorbed, about 5 minutes. Stir in the beans and heat through. Keep the sauce warm.

Meanwhile, bring 6 cups water to a boil. Add salt and stir in the pasta. Continue to stir slowly until the water returns to a boil so the pasta does not stick together. Cook according to the time listed on the pasta package, tasting a minute before in case the package is wrong.

Just before the pasta is cooked, add the basil leaves to the sauce. Drain the pasta and toss with the sauce.

MAKES I SERVING
CALORIES: 590; STARCH: 2 (pasta); VEGETABLES: 2½ (½ pepper, 2 tomatoes);
FAT: 2 (olive oil); FRUIT: 0; DAIRY: 0

Penne with Spinach and Beans

When I am asked to give a cooking demonstration, this is one of my favorite recipes to make because it appeals to a wide audience. And because it comes together so quickly, I can show at least two other pasta dishes. I also like bowtie pasta for this.

2 tablespoons extra virgin olive oil
1 garlic clove, minced (by hand or through a garlic press)
½ to 1 teaspoon dried oregano
2 cups fresh baby spinach or ⅓ cup chopped thawed frozen spinach
Salt and black pepper
½ cup cooked cannellini beans (see page 128), or canned beans, drained
    and rinsed
1 cup canned diced tomatoes
2 ounces whole wheat penne

Heat the olive oil in a medium skillet over medium-low heat. Stir in the garlic and oregano and cook until the garlic is golden, 2 or 3 minutes; do not let it brown. Turn the heat to medium and add the spinach. Season with salt and pepper and cook until wilted, 1 or 2 minutes; frozen spinach should be cooked for 7 minutes, or until any excess water has evaporated. Add the beans and cook until most of the oil is absorbed, 3 to 5 minutes. Stir in the tomatoes and heat through. Reduce the heat and keep the sauce warm.

Meanwhile, bring 6 cups water to a boil. Add salt and stir in the pasta. Continue to stir slowly until the water returns to a boil so the pasta does not stick together. Cook according to the time listed on the pasta package, tasting a minute before in case the package is wrong.

Drain the pasta and toss with the sauce.

MAKES I SERVING
CALORIES: 560; STARCH: 3 (2 pasta, I beans); VEGETABLES: 4 (2 spinach, 2 tomatoes); FAT: 2 (olive oil); FRUIT: 0; DAIRY: 0

Baked Pasta with Chickpeas

Baked pasta dishes are terrific for entertaining because they can be assembled ahead of time and baked when you need them. This is best made with short pasta, such as penne or rigatoni.

2 tablespoons extra virgin olive oil, plus ½ teaspoon to grease the pan
¼ cup red and/or green bell pepper strips or thawed frozen pepper strips
Salt
2 cups baby spinach or ⅓ cup thawed frozen spinach
1 cup canned Italian-style tomatoes
½ cup cooked chickpeas (see page 128), or canned chickpeas, drained
   and rinsed
2 ounces whole wheat pasta

Equipment needed: small (6 × 8-inch) shallow baking dish

Preheat the oven to 350°F. Coat the baking dish with the ½ teaspoon of olive oil. Heat the 2 tablespoons olive oil in a small skillet over medium-low heat. Stir in the pepper, season with salt, and cook until soft, about 10 minutes. Add the spinach, season with salt, and cook until wilted, 1 or 2 minutes; frozen spinach should be cooked for 7 minutes, or until any excess water has evaporated. Transfer the spinach and peppers to a bowl and add the tomatoes and chickpeas. Set aside.

Meanwhile, bring 6 cups water to a boil. Add salt and stir in the pasta. Continue to stir slowly until the water returns to a boil so the pasta does not stick together. Cook slightly less than the time listed on the pasta package to allow for the baking time.

Drain the pasta and add to the bowl with the vegetables and beans. Toss together and scoop into the baking dish. Bake for 20 to 25 minutes, until the top is bubbly.

MAKES 1 SERVING
CALORIES: 605; STARCH: 3 (1 chickpeas, 2 pasta); VEGETABLES: 4½ (½ peppers, 2 each spinach and tomato); FAT: slightly more than 2 (olive oil); FRUIT: 0; DAIRY: 0

Veggie Macaroni and Cheese

Have you noticed lately that "Mac and Cheese" is enjoying a stint on the gourmet food trail? Well, there's no need to avoid it. This healthy version is the answer. Elbow pasta is a good choice but any short pasta is fine.

2 tablespoons extra virgin olive oil, plus ½ teaspoon to grease the dish
¼ cup chopped red onion
Salt and black pepper
1 cup chopped fresh or thawed frozen broccoli
1 tablespoon all-purpose flour
½ cup nonfat or 1% milk, warmed
2 ounces whole wheat pasta
2 tablespoons shredded Cheddar cheese

Equipment needed: small casserole dish

Preheat the oven to 350°F. Coat the casserole with ½ teaspoon of olive oil.

Heat the 2 tablespoons olive oil in a medium skillet over medium-low heat. Stir in the onion, season with salt and pepper, and cook, stirring occasionally, until softened, about 7 minutes. Add the broccoli and cook until slightly softened, about 10 minutes. Sprinkle the flour on the vegetables and mix it in so there is no dry flour visible. Stir in the milk, turn the heat up to medium, and stir constantly with a wire whisk until the sauce thickens.

Meanwhile, bring 6 cups water to a boil. Add salt and stir in the pasta. Continue to stir slowly until the water returns to a boil so the pasta does not stick together. Cook slightly less than the time listed on the pasta package to allow for the baking time.

Drain the pasta and add to the broccoli sauce. Toss to coat the pasta well before spooning into the casserole dish. Sprinkle the cheese over the top. Season with black pepper. Bake until the cheese on top is brown and the casserole is bubbling, 25 to 30 minutes.

MAKES I SERVING

CALORIES: 600; STARCH: 2 (pasta); VEGETABLES: 2½ (½ onion, 2 broccoli);
FAT: slightly more than 2 (olive oil); ; FRUIT: 0; DAIRY: 1½ (½ Cheddar, I milk)

Macaroni with Piquant Tomato Sauce

On a night when my always hungry, grown nephews, Peter and James, paid me an unex-
pected visit, I had to "wing" a meal from what was in my cupboards and freezer. I made
this and then had to repeat it for them on several more of their visits! About ¼ cup
chopped flat-leaf parsley is a nice addition, as is a little grated Romano on top.

2 tablespoons extra virgin olive oil
½ cup chopped red onion (½ small)
½ cup drained jarred hot pepper rings
Salt and black pepper
½ cup Tomato Sauce (page 152) or your favorite meatless jarred sauce
2 tablespoons capers, rinsed
2 ounces whole wheat pasta (any short macaroni shape)

Heat the olive oil in a medium skillet over medium-low heat. Add the
onion and hot peppers, season with salt and pepper, and cook, stirring
occasionally, until the onions are translucent, about 10 minutes. Add the
tomato sauce and capers and heat through; keep warm.

Meanwhile, bring 6 cups water to a boil. Add salt and stir in the pasta.
Continue to stir slowly until the water returns to a boil so the pasta does
not stick together. Cook according to the time listed on the pasta package, tasting a minute before in case the package is wrong.

Drain the pasta and toss with the sauce.

MAKES I SERVING
CALORIES: 515; STARCH: 2 (pasta); VEGETABLES: 3 (I each onion, pepper, and
tomato sauce); FAT: 2 (olive oil); FRUIT: 0; DAIRY: 0

Cheese Ravioli with
Spinach and Red Peppers

I always keep small, whole wheat cheese ravioli in the freezer for a quick meal. I've found that most grocery stores have them in their frozen food sections. The ones I use have 225 calories in 4 ounces. No need to defrost the ravioli before cooking.

2 tablespoons extra virgin olive oil
½ cup diced red onion (½ small)
Salt
2 cups baby spinach or ⅓ cup thawed frozen spinach
½ cup sliced roasted red bell pepper (see page 220) or jarred red pepper
4 ounces frozen small cheese ravioli

Heat the olive oil in a medium skillet over medium-low heat. Stir in the onion, season with salt, and cook, stirring occasionally, until soft, about 5 minutes. Add the spinach and peppers and cook until they absorb almost all the olive oil, about 5 minutes. Keep the sauce warm on low heat.

Meanwhile, bring 6 cups water to a boil. Add salt and stir in the ravioli. Continue to stir slowly until the water returns to a boil so the ravioli do not stick together. Cook according to the time listed on the package, tasting a minute before in case the package is wrong. The ravioli will float on top of the water when they are done or almost done.

Drain the ravioli and transfer to a plate. Top with the vegetables.

MAKES 1 SERVING
CALORIES: 530; STARCH: 3 (ravioli); VEGETABLES: 4 (1 each pepper and onion, 2 spinach);
FAT: 2 (olive oil); FRUIT: 0; DAIRY: ½ (cheese in ravioli)

Pumpkin Ravioli with
Red Onion and Cranberries

This is a nice fall dish—if you can find pumpkin or squash ravioli. They are readily avail-
able in my area but may not be in yours. In that case use cheese ravioli and perhaps
have some roasted pumpkin or winter squash on the side to experience the combina-
tion of flavors.

2 tablespoons extra virgin olive oil
1 cup paper-thin slices of red onion (½ medium)
Salt
2 tablespoons dried cranberries
1 tablespoon all-purpose flour
½ cup low-sodium vegetable broth
3 ounces pumpkin ravioli

Heat the olive oil in a medium pan over medium-low heat. Add the
onion, season with salt, and cook, stirring occasionally, until softened,
about 7 minutes. Stir in the cranberries and cook for 2 to 3 minutes to
heat through. Sprinkle the flour over the onion and use a whisk to completely mix it in so there is no dry flour visible. Pour in the vegetable
broth, raise the heat slightly, and cook until the broth thickens, 3 to 5
minutes. Stir occasionally so it does not stick to the pan.

Meanwhile bring 6 cups water to a boil. Add salt and stir in the ravioli.
Continue to stir slowly until the water returns to a boil so the ravioli do
not stick together. Cook according to the time listed on the pasta package, tasting a minute before in case the package is wrong. The ravioli
will float on top of the water when they are done or almost done.

Drain the ravioli and toss with the sauce.

MAKES I SERVING
CALORIES: 605; STARCH: 2½ (pasta); VEGETABLES: 2½ (½ pumpkin in ravioli, 2 onion);
FAT: 2 (olive oil); FRUIT: 1 (dried cranberries); DAIRY: 0

GOOD TO KNOW

Pasta with Roasted Vegetables

Roasted vegetables make a superb topping for pasta. The high heat of the oven brings out all the sweetness they have to offer. There are three recipes in this chapter that begin with roasting vegetables—Spaghetti with Roasted Bell Peppers and Onions (page 173), Pasta with Roasted Artichokes and Tomatoes (page 174), and Pasta with Creamy Roasted Cauliflower Sauce (page 170). But if you have been preparing your kitchen for Grid Eating, you may have followed the recipe for roasting vegetables on page 214 and have a supply in the freezer. When you freeze them, divide into serving-size portions, and note how much oil is in each serving. Add that amount to the allowances.

Pasta with Creamy Roasted Cauliflower Sauce

This roasted cauliflower gets all dressed up with a cream sauce. Grated Parmesan and black pepper are a nice finish.

1½ cups fresh cauliflower florets or thawed frozen cauliflower
   (about 7 ounces)
2 tablespoons extra virgin olive oil
Salt and black pepper
2 ounces whole wheat pasta (any short shape)
2 bay leaves
½ cup nonfat or 1% milk
2 tablespoons all-purpose flour

Equipment needed: small roasting pan that can also go on top of the stove

Preheat the oven to 425°F. Toss the cauliflower with the olive oil and place in the roasting pan. Season with salt and pepper and roast, stirring two or three times, for 30 to 40 minutes, until the cauliflower is browned.

Bring 6 cups water to a boil. Add salt and stir in the pasta. Continue to stir slowly until the water returns to a boil so the pasta does not stick together. Cook according to the time listed on the pasta package, tasting a minute before in case the package is wrong.

Meanwhile, add the bay leaves to the milk and warm in a small pan on top of the stove, or in the microwave for about 5 minutes on low frequency. Keep warm.

Place the cauliflower pan on the stovetop over medium-low heat. There will be oil in the bottom of the pan. Sprinkle the flour over the cauliflower and toss well so the flour is completely mixed into the oil. There should be no dry flour visible. Remove the bay leaves from the milk and pour the milk into the pan with the cauliflower. Gently stir with a whisk and heat until the sauce thickens.

Drain the pasta and toss with the sauce.

MAKES 1 SERVING
CALORIES: 570; STARCH: 2 (pasta); VEGETABLES: 3 (cauliflower); FAT: 2 (olive oil); FRUIT: 0; DAIRY: 1 (milk)

Pasta with Artichokes and Roasted Red Peppers

My study participants told me that this was one of the recipes they used often for enter-taining. It's a snap to make and quite cheery with its contrasting colors. Of course, it also tastes great.

2 tablespoons extra virgin olive oil
1 cup drained canned artichokes (about half a 14-ounce can)
½ cup sliced roasted red bell pepper (see page 220) or jarred red pepper
Salt
6 large pitted olives, sliced (about 1 ounce)
2 ounces whole wheat pasta (any shape)

Heat the olive oil in a medium skillet over medium-low heat. Stir in the artichokes and peppers, season with salt, and cook for 10 minutes to al-low the vegetables to absorb the flavor of the olive oil. Add the olives and keep warm.

Meanwhile, bring 6 cups water to a boil. Add salt and stir in the pasta. Continue to stir slowly until the water returns to a boil so the pasta does not stick together. Cook according to the time listed on the pasta pack-age, tasting a minute before in case the package is wrong.

Drain the pasta and transfer to the pan with vegetables. Toss together and serve.

MAKES I SERVING
CALORIES: 555; STARCH: 2 (pasta); VEGETABLES: 3 (1 pepper, 2 artichokes);
FAT: 2½ (½ olives, 2 olive oil); FRUIT: 0; DAIRY: 0

Spaghetti with
Roasted Bell Peppers and Onions

The longer the vegetables cook, the sweeter they taste.

½ cup green bell pepper cut into 1 inch-pieces (½ medium)
½ cup red bell pepper cut into 1-inch pieces (½ medium)
½ cup red onion cut into chunks (½ small)
1 garlic clove, cut in half
Salt and black pepper
2 tablespoons extra virgin olive oil
3 or 4 fresh rosemary sprigs (optional)
2 ounces whole wheat pasta

Equipment needed: shallow roasting pan just large enough to hold the vegetables snugly

Preheat the oven to 425°F.

Spread the bell peppers and onion in the baking pan. Add the garlic, season with salt and pepper, and drizzle evenly with the olive oil. Drop in the rosemary and toss to distribute the oil. Roast until the vegetables are partially blackened, 30 to 40 minutes. Stir two or three times as they roast to distribute the olive oil.

When the vegetables have been in the oven for about 15 minutes, bring 6 cups water to a boil. Add salt and stir in the pasta. Continue to stir slowly until the water returns to a boil so the pasta does not stick together. Cook according to the time listed on the pasta package, tasting a minute before in case the package is wrong.

Drain the pasta and toss with the vegetables.

MAKES I SERVING
CALORIES: 460; STARCH: 2 (pasta); VEGETABLES: 3 (I onion, 2 pepper); FAT: 2 (olive oil); FRUIT: 0; DAIRY: 0

Pasta with Roasted Artichokes and Tomatoes

A blissful combination of flavors! If you don't have Italian-style tomatoes, add a minced clove of garlic and a few sprigs of fresh oregano (or 1 teaspoon dried) with the tomatoes.

¾ cup canned artichokes, drained (about half a 14-ounce can)
½ cup chopped red onion (½ small)
Salt and black pepper
2 tablespoons extra virgin olive oil
½ cup canned Italian-style tomatoes
2 ounces whole wheat pasta

Equipment needed: baking pan just large enough to hold the vegetables snugly in one layer

Preheat the oven to 425°F.

Cut the artichokes in half. Combine the artichokes and onion in the baking pan. Season with salt and pepper, drizzle with the olive oil, and toss to distribute. Roast for 35 to 40 minutes, stirring two or three times to distribute the oil, until browned. Add the tomatoes and roast for 15 minutes longer.

When the vegetables have been in the oven for about 30 minutes, bring 6 cups water to a boil. Add salt and stir in the pasta. Continue to stir slowly until the water returns to a boil so the pasta does not stick together. Cook according to the time listed on the pasta package, tasting a minute before in case the package is wrong.

Drain the pasta and toss with the vegetables.

MAKES I SERVING
CALORIES: 500; STARCH: 2 (pasta); VEGETABLES: 3½ (1 each onion and tomato, 1½ artichoke); FAT: 2 (olive oil); FRUIT: 0; DAIRY: 0

Eggplant Lasagna

If you have eggplant already roasted and in your freezer (see page 217), you will find this dish comes together very quickly. No need to thaw the eggplant; just put it in a small frying pan with any oil from the plastic bag or storage container and keep over medium-low heat until it thaws, then turn up the heat and brown. One serving of Roasted Eggplant has 1 tablespoon of olive oil, so cut the amount used here from 2 tablespoons to 1 and the counts will stay the same.

2 tablespoons extra virgin olive oil
2 thick (about ¾-inch) slices eggplant, skin on
Salt and black pepper
1 cup canned diced tomatoes, plain or Italian-style, with juices
¼ cup part-skim ricotta cheese
2 oven-ready (no-boil) whole wheat lasagna noodles, 3 × 7 inches
¼ cup shredded mozzarella cheese (1 ounce)

Equipment needed: small roasting pan and small baking dish just large enough for a lasagna noodle to lay flat

Preheat the oven to 450°F.

Brush 1 tablespoon of the olive oil over both sides of the eggplant slices. Lay in a small roasting pan, season with salt and pepper, and roast for 25 to 30 minutes, until browned. Turn the slices over midway through the cooking. Remove the eggplant from the oven and reduce the oven temperature to 350°F.

Season the tomatoes with salt and pepper and the remaining 1 table-spoon olive oil. Scoop half of them into the baking dish. Spread half the ricotta on one side of one lasagna noodle. Place the noodle, cheese side up, on the tomatoes. Cover the noodle with one slice of eggplant and sprinkle half the mozzarella on top. Spread the remaining ricotta cheese on the second noodle and lay it, cheese side up, in the dish. Top with the other slice of eggplant and the remaining mozzarella. Pour the remaining tomatoes over the top.

Cover with foil and bake for 30 minutes. Remove the foil and bake for 5 to 10 minutes longer, until the cheese is melted.

MAKES I SERVING
CALORIES: 580; STARCH: 2 (lasagna noodles); VEGETABLES: 4 (2 each eggplant and
tomatoes); FAT: 2 (olive oil); FRUIT: 0; DAIRY: 2 (I each ricotta and mozzarella)

Mushroom Lasagna

*Lasagna really is one of cookery's most gratifying and flexible dishes. It can be assem-
bled hours or a day ahead of baking; once baked, it freezes well (freeze serving-size
pieces individually); and there is no end to filling possibilities. I encourage you to be creative with your favorites. Just follow the proportions and technique below and you are on
your way. You can use 2 cups of Tomato Sauce (page 152) in place of the canned tomatoes for a deeper flavor.*

6 tablespoons extra virgin olive oil
1 garlic clove, minced (by hand or through a garlic press)
2 cups chopped red onion (1 medium)
Salt and black pepper
2 cups sliced mushrooms (about 7 ounces)
1 (28-ounce) can crushed tomatoes
1 large egg, beaten
2 cups part-skim ricotta cheese (16 ounces)
½ cup finely chopped fresh flat-leaf parsley
6 ounces oven-ready (no-boil) whole wheat lasagna noodles
6 ounces shredded part-skim mozzarella cheese

Equipment needed: 9 × 13-inch lasagna pan

Heat the olive oil in a large skillet over medium heat. Stir in the garlic
and cook until it is golden but not brown, 2 to 3 minutes. Add the onion,
season lightly with salt and pepper, and cook, stirring occasionally, until
translucent, about 10 minutes. Stir in the mushrooms, season lightly
with salt and pepper, and cook until they begin to color, about 10 minutes. Add the tomatoes, season, and cook for 5 minutes. Remove from
the heat.

Preheat the oven to 350°F.

Stir together the egg, ricotta, and parsley; season with black pepper. Spread half of the cooked vegetables over the bottom of the lasagna pan. Cover with half of the lasagna noodles. Drop half of the ricotta mixture by spoonfuls onto the noodles and then spread it around; a dinner knife works well to do this. Sprinkle on half the mozzarella. Repeat the layers, beginning with the vegetables and ending with the mozzarella.

Bake the lasagna for 30 to 40 minutes, until the cheese on top is browned and the filling is bubbling. The lasagna will cook in less time if the ingredients are warm to start. Let the lasagna sit for 10 to 15 minutes before cutting into 6 equal pieces.

VARIATION: One of my favorite variations is bell pepper lasagna. Use 4 cups of sliced red and green peppers, cooking them slowly in the olive oil until they are very soft (about 20 minutes), and substitute tomato sauce for the chopped tomatoes. Everything else in the recipe stays the same.

MAKES 6 SERVINGS
CALORIES: 495; STARCH: 1 (lasagna noodles); VEGETABLES: 2¾ (⅛ parsley, ⅔ each onion and mushrooms, 1⅓ tomatoes); FAT: 1 (olive oil); FRUIT: 0; DAIRY: 2⅓ (1 mozzarella, 1⅓ ricotta)

Southwest Lasagna

As I wrote before, just about any filling works in lasagna and it doesn't have to be Italian. This black bean, salsa, and Cheddar lasagna is inspired by the flavors of our own Southwest.

6 tablespoons extra virgin olive oil
2 cups chopped red onion (1 medium)
Salt
1¾ cups cooked black beans (see page 128), or canned beans, drained
    and rinsed
1 cup favorite salsa
1 cup canned crushed tomatoes
1 large egg, beaten
1 teaspoon dried oregano or 1 tablespoon chopped fresh oregano

1½ cups part-skim ricotta cheese
6 ounces oven-ready (no boil) whole wheat lasagna noodles
6 ounces shredded Cheddar cheese

Equipment needed: 9 × 13-inch lasagna pan

Preheat the oven to 350°F.

Heat the olive oil in a large skillet over medium-low heat. Stir in the onion, season with salt, and cook, stirring occasionally, until softened but not brown, 7 to 10 minutes. Add the beans and cook for 3 to 5 minutes to heat through. Add the salsa and tomatoes and cook 5 minutes longer, stirring occasionally, to blend the flavors.

Add the egg and oregano to the ricotta and stir until combined and smooth. Pour half of the vegetable and bean mixture into the lasagna pan. Layer half the noodles evenly over the top and spread all the ricotta mixture evenly over the noodles. Top with the remaining noodles and then the remaining vegetables and beans. Sprinkle the Cheddar on top.

Cover with aluminum foil and bake for 20 minutes. Remove the foil and bake 10 minutes longer to brown the top. Let stand for 10 to 15 minutes before serving. Cut into 6 equal pieces.

MAKES 6 SERVINGS
CALORIES: 520; STARCH: 1½ (½ beans, 1 noodles); VEGETABLES: 1 ⅔ (⅓ tomatoes, ⅔ each onion and salsa); FAT: 1 (olive oil); FRUIT: 0; DAIRY: 2 (1 each Cheddar and ricotta)

Southwest Ravioli Casserole

This ravioli casserole has the same great Southwest flavors as the lasagna (page 177) but is quicker to prepare, and it reheats beautifully in the microwave. The ravioli add a nice texture. I've used cheese ravioli here, but vegetable-stuffed ravioli would also be good. Subtract ½ dairy count if using vegetable ravioli.

2 tablespoons extra virgin olive oil
¼ cup chopped red onion
¼ cup sliced green and/or red bell pepper, fresh or frozen
Salt
¼ cup canned black beans, rinsed and drained
½ cup favorite salsa
¼ cup canned crushed tomatoes
3 ounces frozen cheese ravioli
2 tablespoons shredded Cheddar cheese

Equipment needed: small baking dish

Preheat the oven to 350°F.

Heat the olive oil in a medium skillet over medium-low heat. Stir in the onion and bell pepper, season with salt, and cook until beginning to soften, 10 to 12 minutes. Stir in the beans and cook for 3 to 5 minutes to flavor with the oil. Add the salsa and tomatoes and cook until the mixture begins to bubble, 3 to 5 minutes.

Spoon half of the sauce into the baking dish. Place the frozen ravioli over the vegetables. Top with the remaining sauce and sprinkle with the cheese. Cover with aluminum foil and bake for 30 minutes, until the ravioli are tender and the cheese is bubbling.

SERVES I
CALORIES: 585; STARCH: 3 (ravioli); VEGETABLES: 3½ (½ each onion, pepper, and tomato; 2 salsa); FAT: 2 (olive oil); FRUIT: 0; DAIRY: I (½ each Cheddar and cheese in the ravioli)

Vegetable Lo Mein

Many of the women in the Komen study told me that this became their "fast food." It is indeed quick and easy and can be done with many combinations of vegetables.

2 tablespoons extra virgin olive oil
1 cup sliced carrots (2 medium), about ⅛ inch thick
1 cup diced celery (2 medium stalks)
½ cup sliced red onion (½ small)
Salt
⅓ cup reduced-sodium soy sauce
1 teaspoon cornstarch
2 ounces whole wheat spaghetti

Heat the olive oil in a medium pan over medium heat. Stir in the carrots, celery, and onion, season with salt, and cook for 10 minutes. The vegetables will still be somewhat firm.

Add 2 tablespoons cold water to the soy sauce in a small bowl. Whisk in the cornstarch until it is completely blended. Stir into the vegetables and heat until the juices are thickened, stirring occasionally.

Bring 6 cups water to a boil, add salt, and stir in the spaghetti. Continue to stir slowly until the water returns to a boil so the pasta does not stick together. Cook according to the time listed on the pasta package, tasting a minute before in case the package is wrong. Drain the pasta and toss with the cooked vegetables.

VARIATIONS: Try different mixtures of mushrooms, sliced red bell pepper, scallions, and shredded cabbage. Cut the vegetables thin and small enough so they can go in the pan together to cook at the same time. Minced garlic and grated fresh ginger are nice additions. If you want a thinner noodle, use whole wheat capellini or vermicelli.

MAKES 1 SERVING
CALORIES: 510; STARCH: 2 (pasta); VEGETABLES: 5 (1 onion, 2 carrots, 2 celery); FAT: 2 (olive oil); FRUIT: 0; DAIRY: 0

Couscous Pilaf with Spinach and Beans

This is definitely my idea of comfort food. It is a complete meal as made below, but if you want the couscous as a side dish or in place of rice under stews, omit the beans, spinach, and Parmesan. For more flavor, substitute vegetable broth for the water. You can also cook ¼ cup chopped onion in the oil before stirring in the couscous. Any herb is a nice addition.

1 tablespoon plus 2 teaspoons extra virgin olive oil
2 ounces whole wheat couscous (slightly less than 6 tablespoons)
2 cups baby spinach leaves or ⅓ cup thawed frozen spinach
Salt
½ cup cooked cannellini beans (see page 128), or canned beans, drained
    and rinsed
2 tablespoons grated Parmesan cheese

Heat 2 teaspoons of the olive oil in a medium saucepan over medium heat. Add the couscous and stir to coat with the oil. Cook until slightly browned, 2 to 3 minutes. Pour in 1 cup boiling water. Raise the heat slightly until the water returns to a boil, then reduce to medium-low and cook until all the water is absorbed, 10 to 12 minutes.

While the couscous is cooking, heat the remaining 1 tablespoon olive oil in a medium skillet over medium-low heat. Add the spinach, season with salt, and cook until it wilts, 1 or 2 minutes; frozen spinach should be cooked for 7 minutes, or until any excess water has evaporated. Stir in the beans and cook until warmed. Toss with the couscous and serve sprinkled with the Parmesan.

MAKES I SERVING
CALORIES: 570; STARCH: 3 (I beans, 2 couscous); FAT: I ⅔ (olive oil);
VEGETABLES: 2 (spinach); FRUIT: 0; DAIRY: ½ (cheese)

GOOD TO KNOW

Sea Salt

All the recipes in this book that call for salt simply say "salt," but I encourage you to try sea salt. Sea salt and regular old-fashioned table salt basically have the same minerals—they both consist mostly of two chemicals, sodium and chloride. The real difference is not in the health value, but in the taste and texture. Unlike heavily processed table salt, sea salt is produced with a minimum of processing—it is primarily a result of the evaporation of seawater. This leaves the salt with some trace minerals that are insignificant to health but important to flavor and coarseness. I think both these factors lead to using less salt: The coarseness lets you see (or your fingers feel) exactly how much you are using, and adding just a few shakes of sea salt to vegetables as they cook in olive oil is all you need to intensify the flavor.

Table salt does have added iodine, which is needed to make the thyroid hormones. Food companies began to add iodine to salt in the 1920s after finding that goiter, a result of iodine deficiency, was common among young men who lived in the interior/lake regions of the country. The same deficiency was not found in coastal areas since plant products grown near the ocean contain iodine. If you have a concern, use iodized salt for boiling the starch and sea salt for seasoning the sauce.

You will notice that I usually salt each ingredient as it goes into the pan. This is to bring up the flavor of the individual ingredient. Salt lightly, so you don't wind up oversalting the entire dish. If you are trying to control the sodium in your diet, keep in mind that the majority of sodium in the American diet, and the problems associated with it, comes not from the salt you use, but from packaged, commercial, restaurant, and fast foods. If you are making your own meals, you can control how much sodium you are taking in.

Seafood

On the Pink Ribbon Diet, seafood and poultry are limited to 12 ounces per week; not 12 ounces of each category, but 12 ounces total. (Meat can make up 6 ounces per month of that total.) Without fail, when I tell patients or study participants about the limits, they become perplexed and, I won't say they wail, but it is close: "But, where will I get my protein?"

My standard answer is, "If people worried as much about their vegetable intake as they did about protein, they would all be a lot healthier." It's not that seafood and animal protein are necessarily bad for you, it's just that you do not need it and it uses up calories that are better spent on foods that will help decrease your risk of breast cancer.

Are you familiar with those contests in which the winner gets to take a large shopping cart around a store and fill it with anything she likes? Well, let's say you won and the store is a fancy women's clothing emporium. You start out and the first thing you see is a display of the most beautiful pocketbooks ever. So you put a bunch in your cart, then you add more, and you didn't see the large satchels at first so you add a number of those. Next are shoes and you fill up on those; fill up so much that by the time you get to skirts, slacks, blouses, coats, and underwear, you have no more room in the cart. You have a cartful of nice additions to a wardrobe, but not much of a wardrobe. Same is true with food: If you fill up on nice additions to the diet—seafood and poultry, for example—you'll have to go without underwear, and by that I mean needed nutrients.

Perhaps you are wondering about the health benefits of the omega-3 fatty acids that fish contains. This is not as straightforward as you might think. Omega-3 fatty acids are a family of polyunsaturated fats, and that family has members with different characteristics and behaviors—such as how long they are and how many carbons there are in their chain.

Regardless of claims you may have heard, the omega-3's found in canola oil, flaxseed, and walnuts are short-chained and not the type needed to produce hormone-like compounds that our bodies can use. The longer-chain omega-3's, found in fatty fish like Atlantic salmon, mackerel, and tuna, are the ones with health benefits. No study has shown that our bodies can convert short-chained omega-3's into long-chained omega-3's.

So why not just eat a lot of fatty fish? As I said, it is not that straight-forward. Studies show that just adding omega-3's to your diet will not make you healthier—especially if you add fish pills which have not been shown to decrease any disease. Furthermore, studies have failed to prove that omega-3's protect against cancer. What does seem important is the ratio of omega-3's to omega-6's (another family of polyunsaturated fats that are found in vegetable oils like soybean, safflower, and corn oil). When a diet is high in omega-6's and low in omega-3's, the body is in an unbalanced and unhealthy state. The answer is not to add more omega-3's, but to eliminate the sources of omega-6's.

Buying Fish

People who study such things have determined that consumption of fish in restaurants has been steadily climbing in the United States in the last ten years while market sales have remained virtually unchanged. That indicates that we want to eat fish, we just don't know how to buy it and cook it. The confusion is understandable. An array of fish from all over the world now shares room with what used to be a couple of local and familiar varieties. Here is the good news: Almost all varieties have good substitutes. Just pick a fish with similar characteristics, such as firm-fleshed or delicate, lean or oily, thick or thin, full-flavored or mild. Rely on the advice of the fishmonger (if he or she seems knowledgeable) or on a dependable fish cookbook. Mark Bittman's *Fish: The Complete Guide to Buying and Cooking* is especially helpful.

Before you choose your seafood, you have to determine if the place you are about to plop down your dollars is where you want to be. It doesn't have to be a fish market; many grocery stores have top-notch seafood departments. Let your nose lead you. A strong, unpleasant, fishy odor is an alarm to vacate the premises, immediately. Properly caught and stored fresh fish does not smell.

Look at how the seafood is stored. It should be sitting on ice and not piled so high that the top pieces are too far away from the ice to remain chilled. The eyes of whole fish should be clear not cloudy, the gills bright red, and the tail moist with no shriveling. Fish fillets and steaks should look moist, glistening, and translucent. Whole or in pieces, fish flesh should be taut. Borrow a rubber glove from the fishmonger and ask him to hold up the fish. Press your finger into the flesh; there should be no indentation.

Clams and mussels should have tightly closed shells, or close immediately when you tap them. Scallops are usually sold out of their shells and may be listed as "dry" or "wet." Dry are sold in their own juices, which is preferable since the saline bath of "wet" scallops seeps out during cooking and prevents the scallops from browning.

Shrimp, with or without shells, arrives at most markets frozen. Freezing does not interfere with the quality, so I recommend you buy frozen shrimp rather than wonder how long they have been defrosted. Store them in portion sizes in your freezer and thaw them either in the refrigerator, at room temperature, or in a bowl of cold water (tightly ensconced in a plastic bag).

All seafood should stay cold until you are ready to use it. When you get home, double-bag the fish in plastic bags and lay it in a shallow dish or a bowl; cover it with ice cubes and keep refrigerated for a day or two, changing the ice as it melts.

I am not a fan of frozen fish, but many of my patients tell me they have found very good sources of single-serving fish fillets. If fish has been handled properly from the catch to the ice, there is no reason it should not be good. Do not, however, purchase frozen prepared, breaded seafood; it most likely contains more calories than you want for a meal. Canned fish and shellfish—tuna fish, minced clams, and crabmeat, for example—are allowed and convenient since they can always be in your pantry.

Cooking Fish

When it comes to judging the cook time for flat fish, first let me say that whoever came up with "cook until the fish flakes easily" obviously had a deep-seated, pathological loathing for fish. Once fish flakes, it is seriously and sadly overcooked. Fish is cooked when it turns from translucent to opaque; you can check by inserting the tip of a small sharp knife or the tines of a fork into the flesh and gently nudging it apart so you can peek inside. When you insert the knife, cooked fish will offer no resistance, while "almost there" fish will have just the slightest resistance at the center. Fish will continue to cook after it is removed from the heat, so I try to catch it just before that last bit of resistance disappears.

There is a fairly good timing guide for fish: Cook for 10 minutes per inch of thickness, measured at the thickest place. It is only a guide, since a number of factors can alter the equation, such as the density of fish, the heat source, and so on. I count on 8 to 10 minutes per inch and start testing at the lower end.

Cod with Fruit Salsa

For good substitutions, use other white fish such as hake or pollack.

2 tablespoons diced kiwi
2 tablespoons diced fresh pineapple, or canned pineapple in juice, drained
2 tablespoons lime juice
1 tablespoon chopped fresh basil
3-ounce cod fillet, rinsed and patted dry
Salt and black pepper
1 tablespoon extra virgin olive oil
1 tablespoon lemon juice

At least 2 hours before cooking the fish, mix the kiwi, pineapple, 1 tablespoon of the lime juice and the basil in a small bowl. Cover and set aside. The salsa can be made up to the day before, but add the basil about 2 hours before serving.

Season both sides of the cod fillet with salt and pepper. Heat the olive oil in a small pan over medium heat. Add the cod and cook until brown, about 4 minutes. Turn the fillet over and pour the lemon juice and remaining 1 tablespoon lime juice over the fish. Reduce the heat to medium-low and cook until the fish is opaque and the tines of a fork easily pierce the flesh, 2 to 3 minutes, depending on the thickness of the fish.

Serve the fish with the salsa on top.

MAKES I SERVING
CALORIES: 250; STARCH: 0; VEGETABLES: 0; FAT: I (olive oil); FRUIT: ½ (¼ each kiwi and pineapple); DAIRY: 0; SEAFOOD: 3 ounces (cod)

Mediterranean Baked Fish

I particularly like Roasted Potatoes (page 221) with this dish. You can use any white fish. See Cooking Fish, page 185, to learn about timing.

3-ounce cod fillet, rinsed and patted dry
Salt and black pepper
½ cup canned plain or flavored diced tomatoes with juice
¼ cup jarred hot pepper rings
1 tablespoon extra virgin olive oil
6 pitted large olives, sliced
1 tablespoon capers, rinsed and drained

Equipment needed: small baking dish

Preheat the oven to 350°F.

Sprinkle both sides of the fish fillet with salt and pepper and place in the baking dish.

Mix together the tomatoes, hot peppers, olive oil, olives, and capers and season with a little more salt and pepper. Spread the mixture over the fish. Bake for 8 to 10 minutes per inch of thickness, until the tines of a fork easily pierce the flesh and the fish is opaque.

MAKES I SERVING
CALORIES: 280; STARCH: 0; VEGETABLES: 1½ (½ peppers, I tomatoes); FAT: 1½ (½ olives, I olive oil); FRUIT: 0; DAIRY: 0; SEAFOOD: 3 ounces (cod)

Roasted Halibut Steak

You may be used to cooking fish steaks on the grill, but now that you know the potential hazards of grilled seafood protein (see page 38), you won't want to prepare it that way anymore. No need to fret. Roasted fish steaks are delicious. You can roast any variety of fish steak you like, firm-textured or delicate, since you don't have to worry about it falling through the grill rack. In addition to halibut, some of the fish you are apt to find cut into steaks are tuna, swordfish, salmon, shark, and cod. Check the Diet Grid for calorie counts; oily fish will be higher. Use a single herb or make a combination as below, varying the herbs according to what you have and what you like. Use fresh or dried—keeping in mind that 1 teaspoon dried herbs is equivalent to 1 tablespoon fresh.

1 tablespoon extra virgin olive oil
1 teaspoon lemon juice
¼ cup chopped fresh herbs (any combination of parsley, chives, chervil,
    basil) or 4 teaspoons dried
3-ounce halibut steak, about ¾ to 1 inch thick, rinsed and patted dry
Salt and black pepper

Equipment needed: small baking dish

Preheat the oven to 425°F.

Whisk the olive oil with the lemon juice in a shallow dish. Spread the herbs out on a plate. Sprinkle both sides of the halibut steak with salt and pepper. Dip both sides first in the oil, then in the herbs.

Place the steak in the baking dish, Roast, turning the steak once, for 8 to 10 minutes per inch of thickness, until the tines of a fork pierce the fish easily. Serve with a squeeze of lemon juice if desired.

MAKES 1 SERVING
CALORIES: 210; STARCH: 0; VEGETABLES: 0; FAT: 1 (olive oil); FRUIT: 0; DAIRY: 0;
SEAFOOD: 3 ounces (halibut)

Monkfish with Onions, Tomatoes, and Potatoes

Often referred to as "poor man's lobster," monkfish does have a similar flavor, probably because shellfish makes up most of its diet. Also commonly known as "angler fish," it has a firm texture and delicate flavor—as well as an extremely large, ugly head if you have seen the whole fish. Monkfish fillets are actually the tail of the fish (the only part that is eaten). If your fish market has not removed the thin gray membrane that covers the tail, use a small sharp knife and your fingers to take off as much as you can. Monkfish has very dense meat, so it may take longer than 10 minutes per inch to cook.

6 ounces unpeeled red potatoes, scrubbed clean and cut into 1-inch
   pieces
2 tablespoons extra virgin olive oil
Salt and black pepper
½ cup chopped red onion (½ small)
½ cup plain or flavored canned diced tomatoes
3-ounce monkfish fillet, rinsed and patted dry

Equipment needed: small baking dish

Preheat the oven to 425°F.

Put the potatoes in the baking dish and drizzle with 1 tablespoon of the olive oil. Season with salt and pepper and roast for 15 to 20 minutes, until they can be pierced with a fork but are not completely cooked.

Meanwhile, heat the remaining 1 tablespoon olive oil in a small skillet over medium heat. Stir in the onion, season with salt and pepper, and cook, stirring occasionally, until softened, 5 to 7 minutes. Add the tomatoes and bring briefly to a boil. Reduce the heat to medium-low and keep warm on the stove while you cook the fish.

Sprinkle both sides of the monkfish fillet with salt and pepper and place on top of the potatoes. Bake until the fish is opaque, 10 to 12 minutes per inch of thickness. Pour the vegetables over the fish and potatoes and serve.

VARIATION: Because its flavor is so mild, monkfish lends itself to a number of sauces. Here's one favorite: Omit the potatoes and brush the fish on both sides with the olive oil, then roast as above. Meanwhile, after sautéing the onions, stir in ½ cup roasted red peppers—your own or jarred. Transfer the vegetables with their oil to a food processor, add 2 teaspoons balsamic or red wine vinegar, salt, and a pinch of sweet paprika and puree. Pour into a small saucepan and keep warm. Serve the sauce with the fish.

MAKES 1 SERVING
CALORIES: 500; STARCH: 2 (potatoes); VEGETABLES: 2 (1 each onion and tomatoes); FAT: 2 (olive oil); FRUIT: 0; DAIRY: 0; SEAFOOD: 3 ounces (monkfish)

Clams with Spinach, Tomatoes, and Orzo

Don't let the name and shape fool you. "Orzo" is Italian for "barley" and the shape looks like rice but it is really small pasta. Here whole wheat orzo is cooked in a quick, semi-risotto style for a creamy texture.

½ cup canned clams (3 ounces) with 2 tablespoons clam juice
2 tablespoons extra virgin olive oil
2 cups baby spinach or ⅓ cup thawed frozen spinach
Salt
3 ounces whole wheat orzo (9 tablespoons)
½ cup canned plain or Italian-style tomatoes with juice

Stir the clam juice into 1 cup boiling water in a small saucepan. Keep at a bare simmer on the stove.

Heat the olive oil in a medium skillet or saucepan over medium heat. Stir in the spinach, season with salt, and cook until well-wilted and beginning to absorb the oil, about 5 minutes. Add the orzo, stir so it is coated with the oil, and cook until it feels hot to the back of your finger, about 3 minutes. Add the tomatoes and bring to a boil. Add half of the clam juice water. Adjust the heat so the liquid is at a visible simmer but not a furious boil. Cook, stirring occasionally, until most of the liquid is absorbed, 3 to 4 minutes. Add the rest of the clam water and the clams. Cook, stirring occasionally, until the liquid is absorbed and the orzo is tender, about 4 minutes.

MAKES I SERVING
CALORIES: 650; STARCH: 3 (orzo); VEGETABLES: 3 (1 tomato, 2 spinach); FAT: 2 (olive oil); FRUIT: 0; DAIRY: 0; SEAFOOD: 3 ounces (clams)

Clams and Cannellini Beans

Roasted red pepper adds a nice flavor to clams and beans. For a bit of zing, add a pinch or more of hot pepper flakes with the garlic. Serve with a thick toasted slice of whole wheat Italian bread brushed with olive oil.

1 tablespoon extra virgin olive oil
1 garlic clove, minced (by hand or through a garlic press)
½ cup roasted red bell pepper (see page 220) or jarred red pepper
Salt
½ cup cooked cannellini beans (page 128), or canned beans, drained and
    rinsed
½ cup canned clams (3 ounces) with a couple tablespoons clam juice

Heat the olive oil in a small pan over medium-low heat. Stir in the garlic and cook until translucent but not brown, about 5 minutes. Add the red pepper, season with salt, and cook until most of the oil is absorbed, 8 to 10 minutes. Stir in the beans and cook for 5 minutes to blend the flavors. Add the clams and their juice. Increase the heat to medium and cook until most of the liquid has evaporated, about 5 minutes.

MAKES I SERVING
CALORIES: 375; STARCH: 1 (beans); VEGETABLES: 1 (pepper); FAT: 1 (olive oil); FRUIT: 0; DAIRY: 0; SEAFOOD: 3 ounces (clams)

Spicy Clams with Arugula and Roasted Tomatoes

You can enjoy this by itself or spoon it over cooked brown rice or Mashed Potatoes (page 226). Or, tuck a toasted, thick slice of Italian bread (brushed with olive oil, rubbed with garlic, and sprinkled with chopped parsley) into the bowl and let it absorb the flavors.

1 tablespoon extra virgin olive oil
¼ cup jarred hot pepper rings, cut in half
Salt
1 cup arugula
¾ cup Roasted Canned Tomatoes (page 222)
½ cup canned clams (3 ounces) with a little over 1 tablespoon clam juice

Heat the olive oil in a medium skillet over medium heat. Stir in the hot peppers, season with salt, and cook until slightly softened, about 5 minutes. Add the arugula, reduce the heat to medium-low, and cook, stirring occasionally, until the arugula is wilted and has had time to absorb the olive oil, 3 to 5 minutes.

Add the roasted tomatoes and cook until heated through, 3 to 4 minutes. Add the clams and cook until warmed through, 3 to 4 minutes. If the dish has more liquid than you want, turn the heat up to reduce it a bit. If you would like more, add some juice from the clam can or a bit of hot water.

VARIATION: A nice big clove of garlic, minced, is fine cooked with the pepper rings and so is some chopped flat-leaf parsley added just before the arugula.

MAKES I SERVING
CALORIES: 450; STARCH: 0; VEGETABLES: 3 (½ peppers, I arugula, I ½ tomato sauce);
FAT: 2 (olive oil, including tomato sauce); FRUIT: 0; DAIRY: 0; SEAFOOD: 3 ounces (clams)

Clams with Roasted Potatoes
and Tomatoes

This classic Mediterranean combination of clams, potatoes, and tomatoes gets a new twist and added flavor by roasting the potatoes.

6 ounces unpeeled potatoes, scrubbed clean and cut into 1-inch pieces
2 tablespoons extra virgin olive oil
Salt and black pepper
1 cup canned Italian-style tomatoes with juice
½ cup canned clams (3 ounces) with some juice

Preheat the oven to 425°F.

Place the potatoes in the baking dish and drizzle with 1 tablespoon of the olive oil. Season with salt and pepper and roast for 15 minutes, just until they start to brown. The potatoes will not be fully cooked. Stir in the tomatoes and remaining 1 tablespoon olive oil. Return the dish to the oven and continue roasting for about 20 minutes, or until the potatoes are tender, browned, and most of the tomato liquid has evaporated. Scatter the clams and a bit of their juice over the top and roast 5 minutes longer, until the clams are heated through. Use a fork to slightly smash the potatoes so they absorb the juices and serve.

Equipment needed: small baking dish

MAKES I SERVING
CALORIES: 525; STARCH: 2 (potatoes); VEGETABLES: 2 (tomatoes); FAT: 2 (olive oil); FRUIT: 0; DAIRY: 0; SEAFOOD: 3 ounces (clams)

Mussels with Tomato-Spinach Sauce

This makes a fine lunch but is even better—really delicious—served over cooked capellini or vermicelli (2 ounces); add 2 starches and about 210 calories. The pasta absorbs all the flavors so the taste goes on and on. Many of the mussels on the market are farm-raised, so they are pretty clean when you buy them, with beards and barnacles removed. If yours aren't, remove them before cooking.

1 tablespoon extra virgin olive oil
1 garlic clove, minced (by hand or through a garlic press)
2 cups baby spinach or ⅓ cup thawed frozen spinach
Salt and black pepper
½ cup canned plain or flavored crushed tomatoes
10 to 12 mussels (about 3 ounces meat) scrubbed clean, beards and
    barnacles removed
¼ cup fresh basil leaves, torn into small pieces

Heat the olive oil in a medium pan over medium heat. Add the garlic and cook until just translucent but not browned, 2 to 3 minutes. Add the spinach, season with salt, and cook until wilted, 1 to 2 minutes; frozen spinach should be cooked for 7 minutes, or until any excess water has evaporated. Stir in the tomatoes, season with salt and pepper, and increase the heat to medium. Add the mussels, cover, and cook until the shells open, 3 to 5 minutes. Discard any mussels that do not open. Stir in the basil and serve.

MAKES I SERVING
CALORIES: 275; STARCH: 0; VEGETABLES: 3 (1 tomatoes, 2 spinach); FAT: 1 (olive oil);
FRUIT: 0; DAIRY: 0; SEAFOOD: 3 ounces (mussels)

Crab-Stuffed Mushroom

This makes a nice lunch or, if you cut the olive oil to 1 tablespoon, you can serve as an appetizer before a dinner meal.

3-ounce portobello mushroom cap
2 tablespoons plus 1 teaspoon extra virgin olive oil
¼ cup chopped red onion
¼ cup chopped celery
Salt and black pepper
¼ cup chopped fresh flat-leaf parsley
¼ cup (1 ounce) whole wheat bread crumbs
3 ounces crabmeat, picked over to remove any shells

Equipment needed: small baking dish

Preheat the oven to 350°F.

Use a mushroom brush or damp paper towel to wipe clean the outside of the mushroom cap. Pull out the inside of the mushroom and put the pieces in a small bowl; set aside. Brush or rub 1 teaspoon of the olive oil over the cap and place in the baking dish, rounded cap down.

Heat the remaining 2 tablespoons olive oil in a small pan over medium heat. Add the onion and celery and season with salt and pepper. Cook, stirring occasionally, until some of the oil is absorbed but the vegetables are still firm, about 5 minutes. Stir in the mushroom pieces and cook until most of the oil is absorbed, about 5 minutes longer. Add the parsley, bread crumbs, and crab and combine well.

Mound the stuffing in the mushroom cap. Cover the dish loosely with foil. Bake for 20 minutes, or until the filling is hot.

MAKES 1 SERVING
CALORIES: 440; STARCH: 1 (bread crumbs); VEGETABLES: 3½ (½ each onion, celery, and parsley; 1 mushroom); FAT: slightly more than 2 (olive oil); FRUIT: 0; DAIRY: 0; SEAFOOD: 3 ounces (crab)

Crab with Roasted Red Peppers and Olives

Brown rice is a nice accompaniment to this. For the crabmeat, canned is fine. In addition, grocery stores and fishmongers usually carry crabmeat out of the shells as well as pieces that have been cooked but remain in their shells. Use what is convenient for you.

1 tablespoon extra virgin olive oil
½ cup chopped roasted red bell pepper (see page 220)
    or jarred red pepper
Salt
6 pitted large black olives, sliced
3 ounces crabmeat, picked over to remove any shell pieces

Heat the olive oil in a small pan over medium heat. Add the red pepper, season with salt, and cook, stirring occasionally, until most of the oil is absorbed, about 5 minutes. Stir in the olives and crabmeat. Cook 3 to 4 minutes to heat through.

MAKES I SERVING
CALORIES: 265; STARCH: 0; VEGETABLES: I (pepper); FAT: I ½ (½ olives, I olive oil); FRUIT: 0;
DAIRY: 0; SEAFOOD: 3 ounces (crab)

Spicy Mediterranean Crab

There is enough liquid in this to serve over pasta or rice. Use either canned crab or pieces from the market. If you like your food very spicy, three-alarm hot, increase the amount of hot peppers at will.

1 tablespoon extra virgin olive oil
1 garlic clove, minced (by hand or through a garlic press)
¼ cup jarred hot pepper rings, chopped
Salt
1 cup canned diced tomatoes with some juice
6 pitted large olives, sliced in half

1 tablespoon capers, rinsed and drained

3 ounces crabmeat, picked over to remove any shell pieces

Heat the olive oil in a small pan over medium heat. Add the garlic and cook just until it is translucent but not browned, 2 to 3 minutes. Add the hot peppers, season with salt, and cook, stirring occasionally, until softened, 8 to 10 minutes. Stir in the tomatoes, olives, and capers and cook for 5 minutes to heat through and blend the flavors. Add the crabmeat and cook until heated through, 3 to 4 minutes.

MAKES I SERVING
CALORIES: 300; STARCH: 0; VEGETABLES: 2½ (½ pepper, 2 tomato); FAT: 1½ (½ olives, 1 oil); FRUIT: 0; DAIRY: 0; SEAFOOD: 3 ounces (crab)

Shrimp Stew

This is a great dish to make on a night when you have little time to cook, and it also makes a terrific company meal. Increase the ingredients accordingly, and make the recipe up until the point when the shrimp is added. Five minutes before the dinner bell, add the fish, then the basil and rice, and you have dinner.

2 tablespoons extra virgin olive oil

½ cup finely chopped red onion (½ small)

Salt and black pepper

½ cup strips of red and/or green bell pepper, fresh or thawed frozen
  (½ medium fresh)

1 cup canned plain or flavored diced tomatoes

3 ounces peeled shrimp, rinsed and patted dried

¼ cup fresh basil leaves

1 cup cooked brown rice (see page 138)

Heat the olive oil in a medium pan over medium heat. Add the onion, season with salt, and cook, stirring occasionally, until soft, about 7 minutes. Stir in the bell peppers, season with salt, and cook, stirring occasionally, until tender, 8 to 10 minutes. Add the tomatoes. Season with salt and pepper and bring to a boil. Drop in the shrimp and cook just until they are firm and pink, 3 to 5 minutes. Add the basil and rice.

VARIATION: If you'd like a little Cajun spice, cook one minced garlic clove and ¼ cup chopped celery with the onions until the celery is soft. Before adding the peppers, stir a large pinch of cayenne pepper and ¼ teaspoon dried oregano into the oil. Stir them around for 30 seconds to release the flavors and then finish the recipe as described.

MAKES I SERVING

CALORIES: 610; STARCH: 2; VEGETABLES: 4¼ (¼ basil, I each onion and peppers, 2 tomatoes); FAT: 2 (olive oil); FRUIT: 0; DAIRY: 0; SEAFOOD: 3 ounces (shrimp)

Shrimp with Artichokes and Capers

If you have the starch allowance, serve this over brown rice or pasta.

2 tablespoons extra virgin olive oil
1 garlic clove, minced (by hand or put through a garlic press)
1 cup drained canned artichokes cut into quarters
    (about half a 14-ounce can)
Salt
1 tablespoon capers, rinsed and drained
3 ounces peeled shrimp, rinsed and patted dry

Heat the olive oil in a medium pan over medium-low heat. Add the garlic and cook until just translucent but not browned, 2 to 3 minutes. Stir in the artichokes, season with salt, and cook until they have absorbed most of the oil, 8 to 10 minutes. Add the capers and shrimp and cook until the shrimp is firm and pink, 3 to 5 minutes.

MAKES I SERVING

CALORIES: 395; STARCH: 0; VEGETABLES: 2 (artichokes); FAT: 2 (olive oil); FRUIT: 0; DAIRY: 0; SEAFOOD: 3 ounces (shrimp)

Shrimp and Vegetable Fried Rice

Baby shrimp are best for this recipe. If you can't find them, cut larger shrimp into ¼-inch pieces.

2 tablespoons olive oil
3 ounces peeled shrimp, rinsed and patted dry
Salt and black pepper
1 medium garlic clove, minced
1 tablespoon minced fresh ginger
2 cups spinach or kale, washed, stems and ribs removed, finely shredded
1 cup chopped scallions, white and green parts reserved separately
    (1 bunch)
½ cup grated carrot (1 medium)
½ cup finely chopped celery (1 stalk)
½ cup cooked brown rice (see page 138)
1 scant tablespoon reduced-sodium soy sauce

Heat 1 tablespoon of the oil in a wok or large skillet over high heat until the oil is hot enough to sizzle when a bit of water is splashed in the pan. Add the shrimp, season with salt and pepper, and stir-fry just until firm and pink. Transfer to a side dish with a slotted spoon.

Add the remaining 1 tablespoon oil, garlic, and ginger to the pan and stir-fry briefly to release their flavors; do not let brown. Add the spinach, scallions (white parts only), carrot, and celery. Season lightly with salt and pepper and stir-fry until the vegetables are tender, about 4 minutes.

Toss in the rice and the shrimp and continue to stir-fry just until the rice is heated through. Stir in the soy sauce and check seasonings. Serve immediately, garnished with the chopped scallion greens.

MAKES I SERVING
CALORIES: 565; STARCH: I (rice); FAT: 2 (olive oil); VEGETABLES: 6 (I each carrots and celery, 2 each spinach and scallions); FRUIT: 0; DAIRY: 0; SEAFOOD: 3 ounces (shrimp)

Shrimp with Chickpeas, Feta, and Tomato Sauce

Good Mediterranean flavors that are delicious served over thin spaghetti.

2 tablespoons extra virgin olive oil
¼ cup cooked chickpeas (see page 128), or canned chickpeas, rinsed and
    drained
1 cup canned diced tomatoes
Salt
3 ounces peeled shrimp, rinsed and patted dry
2 tablespoons (½ ounce) feta cheese
2 tablespoons fresh basil leaves
1 tablespoon lemon juice

Heat the olive oil in a small pan over medium-low heat. Add the chickpeas and cook, stirring occasionally, to flavor them with the oil, 5 minutes. Stir in the tomatoes, season with salt, and cook until the tomatoes start to boil, about 5 minutes. Add the shrimp and feta and cook until the shrimp is firm and pink and the cheese has melted, 3 to 5 minutes. Stir in the basil and lemon, heat briefly, and serve.

MAKES 1 SERVING
CALORIES: 440; STARCH: ¼ (chickpeas); VEGETABLES: 2 (tomatoes); FAT: 2 (olive oil);
FRUIT: 0; DAIRY: ½ (cheese); SEAFOOD: 3 ounces (shrimp)

Poultry

If you've studied the Diet Grid, you know that meat (beef, lamb, pork) is allowed as well as poultry. So where, you are wondering, are the recipes for meat? There are none. Although it is allowed, I don't encourage you to eat meat; and if you do, you should have no more than 6 ounces a *month,* and that must be counted as part of your total 12-ounce weekly allowance in this food category. I know you will lose more weight and feel better if you do not include meat in your diet. I allow it because I recognize that some of you, as did some of the women in the program, feel you cannot diet at all if you have to give up meat. I'm also betting on the fact that you, like the women in the program, will soon find that you don't miss it at all and will choose to ignore it. So, there are no red meat recipes.

There *are* very good recipes for chicken and turkey, but let me be clear: You do not *need* either for your health. They are optional.

Poultry skin contains a considerable amount of fat, so if you do include poultry in your diet, it should be skinless. It is very easy today to find skinless cuts—not just breasts—of both chicken and turkey. If you buy a bulk "family pack" of poultry that still has the skin on, remove it from the pieces you plan to cook for yourself or ask the meat department of the grocery store to do it for you.

At first glance, it looks as though chicken breasts are an easy choice for a quick, healthy meal, but you have to be careful. Some breasts can weigh up to 9 ounces, which would use up three-quarters of your weekly allowance. Look for ones that weigh 4 to 6 ounces instead. If all you can find are large ones, cut them up yourself at home. Remove the tender—one piece—then lay the rest of the breast on the counter and with a sharp knife, cut the breast horizontally into two pieces. The breast meat can also be "julienned," that is, cut into small, thin strips, and then stir-fried. For example, you could substitute 3 ounces of cut-up chicken for the shrimp in Shrimp and Vegetable Fried Rice (page 199).

Chicken thighs are sold in a convenient size for this diet; with the skin and bones removed, most weigh just about 3 ounces (6 ounces is the maximum amount of poultry you should consume at one meal in order to keep calories under control for the day). Nancy prefers poultry thighs to poultry breasts because they have more flavor and remain juicier after cooking, especially when they are cooked on the bone. Turkey thighs are the perfect cut to use when you want large chunks of meat for stews. Many markets now sell "cutlets" or very thin "scaloppine" that they slice from the turkey breast. The thin ones are a good substitute in recipes that call for veal or chicken scaloppine. Ground turkey is such a stellar surrogate for ground beef that many fine restaurants use it for burgers, meat loaf, chili, and such with great success.

To calculate the weight of poultry with the bones and skin; consider that roughly 65 percent of the weight is what you eat and the rest, 35 percent, is waste. For chicken with the bone but no skin, count 75 percent as meat and 25 percent as waste. To determine the weight of poultry pieces that are packaged, count them through the wrapping, look at the weight listed on the label, and divide the number of pieces into the total weight of the package.

When you are away from home, at a restaurant or someone else's home for a meal, you can "guesstimate" the ounces of poultry, meat, or seafood by imagining a deck of cards or a hockey puck. That's how big 3 to 4 ounces is—stop eating there!

As always, handle raw poultry carefully. Thoroughly clean all surfaces and utensils the poultry touches and cook all pieces completely before eating.

Chicken Cacciatore

Try to purchase chicken thighs that are all the same size so they cook evenly. The ones we used for this recipe were each 4 ounces. If your market has a meat counter with a real live person cutting meat, ask for the size you want.

8 skinless bone-in chicken thighs (about 2 pounds)

Salt and black pepper

¼ cup extra virgin olive oil

4 cups roughly chopped yellow onions (2 large)

3 cups ½-inch-wide strips of red and green bell peppers
  (about 1½ pounds)

4 cups thinly sliced mushrooms (about 12 ounces)

1 large garlic clove, thinly sliced

½ cup dry red wine (optional)

1 (28-ounce) can Italian plum tomatoes, drained and chopped

Rinse the chicken under cold running water and use paper towels to dry thoroughly. Season on both sides with salt and pepper. Heat the oil over medium heat in a pan that is large enough to hold all the ingredients in one slightly overlapping layer. Add the chicken and cook until lightly browned on both sides; work in batches if necessary so there is room between the pieces; otherwise they will not brown. Transfer the browned pieces to a plate and set aside.

Reduce the heat slightly and add the onions to the pan, season with salt and pepper, and cook until translucent, about 10 minutes. Stir in the bell peppers, mushrooms, and garlic and cook, turning over often, until the mushrooms release their juices, about 10 minutes. Raise the heat to high, pour in the wine if using, and let it reduce to a few tablespoons. Add the tomatoes and bring to a boil.

Return the chicken to the pan, tucking it under the vegetables so pieces are submerged in the pan juices. Cover and reduce the heat. Simmer until the chicken is cooked through and the peppers are tender, about 15 minutes,

MAKES 4 SERVINGS, 2 PIECES CHICKEN EACH

CALORIES: 470; STARCH: 0; FAT: 1 (olive oil); VEGETABLES: 6½ (1 tomatoes, 1½ peppers, 2 onions, 2 mushrooms); FRUIT: 0; DAIRY: 0; POULTRY: about 6 ounces

Turkey Scaloppine with Mushrooms and Artichokes

Mushrooms and artichokes really stretch a recipe, so the 3-ounce serving here seems like more. If you actually want more, cook four scaloppine—6 ounces total; no need to increase any of the other ingredients, just add 135 calories. If you don't see turkey scaloppine (very thin slices of turkey breast) with the packaged poultry, ask the meat cutter behind the counter to slice pieces no more than ⅛ inch thick. Otherwise, substitute the same amount of thin chicken breasts or use chicken tenders.

2 thinly sliced turkey scaloppine (3 ounces total)
Salt and black pepper
2 tablespoons extra virgin olive oil
1 large garlic clove, thinly sliced
1 cup thinly sliced mushrooms (about 3 ounces)
½ cup canned artichoke hearts (in water), each cut into 8 wedges (3 artichokes)
Juice of ½ lemon

Rinse the turkey under running cold water and pat dry with paper towels. Season both sides of each piece with salt and pepper. Heat the oil in a medium pan over medium-high heat. Add the turkey and cook, turning once, until cooked through, about 1 minute per side. Transfer to a plate and set aside.

Reduce the heat to low and add the garlic to the pan. Slowly cook until golden, about 5 minutes. Add the mushrooms, season lightly with salt, and raise the heat slightly. Cook until the mushroom juices release into the pan and then evaporate. Stir in the artichokes and cook, turning often, until heated through. Return the turkey to the pan and add the lemon juice. Heat for a minute and serve.

MAKES I SERVING
CALORIES: 380; STARCH: 0; FAT: 2 (olive oil); VEGETABLES: 3 (I artichoke, 2 mushrooms); FRUIT: less than I (lemon); DAIRY: 0; POULTRY: 3 ounces (turkey)

Aromatic Chicken and Squash Stew

With a few simple changes, you can turn this aromatic chicken dish into an equally fla-
vorful Aromatic Vegetarian Stew. Just omit the chicken and use vegetable broth in place
of the chicken broth; the calories per serving will change to 375. You can also substitute
turkey for the chicken or use chicken breasts.

¼ cup extra virgin olive
1 cup finely chopped yellow onion (1 medium)
3 large garlic cloves, minced
1 jalapeño pepper, seeded and minced
1 tablespoon mild paprika
1 teaspoon red pepper flakes (optional)
1 teaspoon ground cardamom
½ teaspoon ground cinnamon
2 teaspoons salt
4 cups ½-inch-diced peeled butternut squash (about 1½ pounds)
Black pepper
½ cup red wine (optional)
2 tablespoons tomato paste
2 cups low-sodium chicken or low-sodium vegetable broth
1¾ cups cooked chickpeas (see page 128), or canned chickpeas,
    rinsed and drained
12 ounces skinless boneless chicken thighs, cut into ¾-inch pieces
Juice of 1 lemon
⅓ cup chopped fresh flat-leaf parsley

Heat the olive oil in a large soup pot or Dutch oven over medium-low heat. Add the onion, garlic, jalapeño, paprika, pepper flakes (if using), cardamom, cinnamon, and 1 teaspoon of the salt and stir to mix well. Cook for 6 minutes to blend the flavors. Add the squash, turning it over and around so it is well-coated with the oil and spices. Sprinkle with the remaining 1 teaspoon salt and a few grindings of black pepper. Cover and cook gently until the squash is just beginning to soften, 8 to 12 minutes. Check the pan from time to time to be sure the squash is not sticking to the pan; add few tablespoons warm water if necessary.

Uncover, turn up the heat, and pour in the wine if using. Let the wine boil and reduce to a few tablespoons. Stir in the tomato paste, broth, and beans. Reduce the heat, cover, and simmer until the squash is cooked through, 8 to 10 minutes.

Submerge the chicken pieces in the stew, cover, and simmer until the chicken is cooked through, about 5 minutes. Stir in the lemon juice and parsley and serve. Garnish with more parsley if desired.

VARIATIONS: If you want a less "aromatic" stew, or want to play with your own seasonings, omit the jalapeño, paprika, pepper flakes, cardamom, and cinnamon—it still makes a great dish. The red wine adds a nice bit of acid, but it is optional. In other words, the basic technique for a chicken stew is there; make it your own. Whatever your choices, serve the stew over brown rice, if desired, and be sure and count the rice for a starch. The stew, without rice, freezes beautifully.

MAKES 4 SERVINGS, ABOUT 2 CUPS EACH
CALORIES: 450; STARCH: slightly more than ¾ (beans); FAT: 1 (olive oil); VEGETABLES: slightly more than 2½ (less than ½ tomato paste and parsley, ½ onion, 2 squash); FRUIT: 0; DAIRY: 0; POULTRY: 3 ounces

Curried Turkey and Pumpkin Stew

When you are buying pumpkin, don't get carried away and purchase a jack-o'-lantern! Pumpkins suitable for eating are small and usually labeled "sugar pumpkin." Turkey breast works in the recipe, but is not as tasty as thighs.

4 cups chopped peeled sugar pumpkin (1 small, about 2 pounds;
   see below)
½ cup olive oil
2 cups coarsely chopped red onion (1 medium)
3 garlic cloves, coarsely chopped
Salt and black pepper
2 tablespoons curry powder
2 teaspoons paprika

½ ounce fresh ginger, cut into 12 nickel-size pieces

1½ cups chopped fresh or canned tomatoes

1 cup low-sodium chicken broth

1 tablespoon tomato paste

1 pound skinless boneless turkey or chicken thighs, cut into ¾-inch
cubes

½ cup raisins (optional)

To prepare the pumpkin, place on a cutting board and remove the peel by running a sharp knife from stem to root end just deep enough to remove the outer layer. Cut the pumpkin in half, scoop out the seeds, and cut the pulp into ¾-inch pieces.

Heat the olive oil over medium-low heat in large straight-sided pan; a 12-inch pan is a perfect size. Stir in the onion and garlic, season with salt, and cook until the onion is translucent, about 10 minutes; do not let it brown. Sprinkle with the curry powder, paprika, and black pepper and stir in the ginger. Cook for 2 minutes to blend the flavors.

Add the pumpkin and stir to coat. Add a little more salt, cover, and cook until the pumpkin is just barely tender, about 15 minutes.

Stir in the tomatoes, broth, and tomato paste. Bring to a boil and add the turkey, pushing it down into the broth. Reduce the heat to low and add the raisins if using. Cover and simmer until the turkey is cooked through, 8 to 10 minutes.

VARIATIONS: You can make this a vegetarian stew by omitting the turkey and substituting low-sodium vegetable broth for the chicken broth; the calories will change to 320 per serving without the raisins, or 375 with the raisins. For a heartier vegetarian stew, substitute 2 cups drained and rinsed canned red beans for the turkey; the calorie count will change to 422. Either way, serve over brown rice or barley.

MAKES 4 SERVINGS, ABOUT 2 CUPS EACH

CALORIES: 460 without raisins, 515 with raisins; STARCH: 0; FAT: 2 (olive oil); VEGETABLES: 3¾ (¾ tomatoes, 1 onion, 2 pumpkin); FRUIT: 0 without raisins, 1 with raisins; DAIRY: 0; POULTRY: 4 ounces

Italian Turkey Meatballs

I think even your Italian grandmother—if you have one—would approve of these turkey meatballs. They taste as they should! The meatballs are dropped into the hot tomato sauce without any prior browning so they remain very tender. If you prefer a little crustiness, you can bake them before adding them to the sauce: Arrange in one layer in a baking pan and bake in a 350°F oven for 10 minutes, turning over a few times. If you want to bake the meatballs until they are completely done (instead of cooking them in the sauce) they will take about 25 minutes total.

¼ cup extra virgin olive oil
½ cup minced yellow onion (1 small)
2 to 4 large garlic cloves, minced
Salt and black pepper
2 cups finely chopped escarole or romaine lettuce
2 tablespoons finely chopped fresh flat-leaf parsley or basil
1 pound ground turkey
½ cup dry old-fashioned (rolled) oats
¼ cup grated Parmesan cheese (optional)
1 large egg, beaten
3 cups Tomato Sauce (page 152) or your favorite jarred meatless sauce

Heat the olive oil over medium-low heat in a pan large enough to hold the vegetables. Add the onion and garlic, season with a pinch of the salt, and cook until the onion is softened but not browned, about 7 minutes. Add the escarole and parsley, season with a pinch more salt and some pepper, and cook until the escarole is completely tender and has absorbed almost all the oil, 8 to 10 minutes. If there is excess water in the pan from the escarole, turn up the heat and boil it away. Let the vegetables cool to room temperature.

Combine the cooked vegetables and turkey in a mixing bowl and add the oatmeal, cheese (if using), egg, and salt. Use a fork or your hands to mix the ingredients until thoroughly combined. Form the mixture into 12 meatballs; wet your hands with cold water to make forming the meatballs easier.

Bring the tomato sauce to a boil in a large saucepan over medium-high heat. Drop in the meatballs, reduce the heat, and simmer until the meatballs are cooked through, about 20 minutes.

MAKES 6 SERVINGS, 2 MEATBALLS EACH
CALORIES: 350 without cheese, 365 with cheese; STARCH: less than 1 (oats); FAT: ⅔ (olive oil); VEGETABLES: about ½ (1/6 onion, ⅓ escarole); FRUIT: 0; DAIRY: 0 without cheese, ⅔ with cheese; POULTRY: 2 ⅔ ounces (turkey)

Old-Fashioned Meat Loaf

This is a great family dish, but if you are the only one who will be eating it, cut the finished meat loaf into six even pieces. Wrap what you won't eat within a few days in individual servings and freeze. Or you can bake the meat mixture in a 12-cup muffin pan. It will take about 25 minutes, and two meat "muffins" are equivalent to one slice of meat loaf.

¼ cup extra virgin olive oil, plus a little more for the pan
1 cup finely chopped yellow onion (1 medium)
2 large garlic cloves, minced
Salt
1 cup grated carrots (2 medium)
2 cups baby spinach or ⅓ cup thawed frozen chopped spinach
1½ pounds ground turkey
1 cup Tomato Sauce (page 152) or jarred meatless tomato sauce
1 large egg, beaten
½ cup dry old-fashioned (rolled) oats
2 teaspoons Italian seasoning
Black pepper

Equipment needed: 9 × 5 × 3-inch loaf pan

Preheat the oven to 375°F. Lightly oil the bottom and sides of the loaf pan.

Heat the oil over medium-low in a pan large enough to hold all the vegetables. Add the onion and garlic, season lightly with salt, and cook slowly until softened, about 7 minutes. Stir in the carrots, salt lightly, and cook until beginning to soften, 3 to 4 minutes. Add the spinach, salt lightly, and cook until completely wilted and almost all the oil has been absorbed. Let cool.

Mix together the cooked vegetables and turkey in a mixing bowl with a fork or your hands. Add ½ cup of the tomato sauce, the egg, oats, seasoning, and black pepper and mix just until combined. Pat into the prepared pan and cover with remaining sauce. Bake for 1 hour and 15 minutes, or until an instant-read meat thermometer registers 160°F when inserted into the center of the loaf.

MAKE 6 SERVINGS
CALORIES: 335; STARCH: less than I (oats); FAT: ⅔ (olive oil); VEGETABLES: I ⅓ (⅓ each carrots, onion, spinach, and tomato sauce); FRUIT: 0; DAIRY: 0; POULTRY: 4 ounces (ground turkey)

South of the Border Meat Loaf

A nice variation of Old-Fashioned Meat Loaf (page 209). Mashed Potatoes (page 226) are awfully nice on the side of any meat loaf.

¼ cup extra virgin olive oil, plus a little for the pan
1 cup finely chopped yellow onion (1 medium)
1 cup chopped red bell pepper (1 medium)
¼ cup finely chopped jalapeño pepper (1 pepper)
2 large garlic cloves, minced
1 teaspoon chili powder
Salt
2 cups fresh spinach or ⅓ cup thawed frozen chopped spinach
¼ cup chopped fresh cilantro
1½ pounds ground turkey
1 cup favorite salsa
1 large egg, beaten
½ cup dry old-fashioned (rolled) oats

Equipment needed: 9 × 5 × 3-inch loaf pan

Preheat the oven to 375°F. Lightly oil the bottom and sides of the loaf pan.

Heat the oil over medium-low heat in a pan large enough to hold all the vegetables. Add the onion, bell pepper, jalapeño pepper, garlic, and chili powder. Season lightly with salt and cook slowly until the vegetables are softened, about 10 minutes. Add the spinach and cilantro and turn around in the oil to coat. Season with another small amount of salt and cook until the spinach is completely wilted and most of the oil has been absorbed, about 7 minutes. Let the vegetables cool.

Mix together the cooked vegetables and turkey in a mixing bowl with a fork or your hands. Add ½ cup of the salsa, the egg, oats, and 1 teaspoon salt. Mix until evenly distributed. Pat the mixture into the loaf pan and cover with the remaining salsa. Bake for 1 hour to 1 hour and 15 minutes, until an instant-read thermometer registers 160°F when inserted into the center of the loaf.

MAKES 6 SERVINGS
CALORIES: 335; STARCH: less than 1; FAT: ⅔ (olive oil); VEGETABLES: 2 (⅙ pepper and parsley, ⅓ each onion and pepper, ½ spinach, ⅔ salsa); FRUIT: 0; DAIRY: 0; POULTRY: 4 ounces (turkey)

10

Vegetables, Side Dishes, Casseroles, and Extras

Vegetables, vegetables, vegetables. Nancy told me about a commercial she saw some years ago on TV. A little boy is sitting at the kitchen table staring quizzically at the label on a Libby's can that reads "Libby's. Libby's. Libby's." Finally the boy asks his mother why Libby's put its name on the can three times. The mother tells him it's because they think so highly of their product. After a moment of thought, the boy then asks, "Then why don't you call me 'Stanley, Stanley, Stanley'?" Well, I think highly enough of vegetables that if the computer didn't keep warning of "repeated word" I'd fill the page with "vegetables." I'm not alone in this. Michael Pollan opens his well-researched book *In Defense of Food* with the pithy advice, "Eat Food. Not too much. Mostly plants." He explains his recommendation by summarizing his polling of experts who study the effects of diet on health: "Scientists may disagree about what's so good about eating plants . . . but they do agree that plants are probably really good for you, and certainly can't hurt. In all my interviews with nutrition experts, the benefits of a plant-based diet provided the only universal consensus."[1]

Four servings a day, a total of 2 cups—that's the minimum you should eat. But that's just the minimum. You should eat more. Vegetable allowances are unlimited—as long as you remember to count any olive oil used for cooking them so you don't exceed the daily allowance of fat. Eat vegetables religiously at lunch and dinner. Fit veggies into breakfast by cooking them with an egg.

For some of you the 2-cup minimum will offer no problem. Others of you will stare into the vegetable bin, perplexed, wondering how on earth you can do it. You are the ones who usually say that you just don't like vegetables—well maybe canned peas, but nothing else. I don't believe that vegetables are at fault. It's the cooking method. Most Americans steam, boil, or microwave their vegetables with little or no seasoning so they have no appeal to someone who doesn't simply love the taste of vegetables. I am always amazed at how many vegetable haters (especially kids) are converted to lovers when they try them cooked in olive oil with a bit of salt.

All vegetables are good for you, but certain ones are more likely to decrease the risk of breast cancer, such as the whole cruciferous or cabbage family—broccoli, Brussels sprouts, cabbage, cauliflower, kale, kohlrabi, and Swiss chard; also, any vegetable with deep color has carotenoids that will fight cancer (see page 241) and you should incorporate those into your diet often. If preparing fresh vegetables seems too much of a chore, use frozen ones. They are all ready to use and, because they have no waste, are usually a very good buy. Although they are seldom good just defrosted, they cook up very well in olive oil. Fresh or frozen, eat a variety: Try ones you've never eaten before or those you haven't eaten in a long time. You might surprise yourself, especially if you roast vegetables, which gives them an entirely different (and often sweeter) taste than when cooked on the stovetop. Even those who claim to have never met a vegetable they like change their minds when they are introduced to roasted vegetables. The first recipe in this chapter is a general guide to roasting all vegetables.

Roasting Vegetables

Roasting intensifies the flavors and brings out the sweetness of vegetables. The longer you roast them, the browner, sweeter, and crispier they become. Every vegetable can be roasted—alone or in combination with others.

HOW TO

TEMPERATURE The oven temperature for roasting should be at least 425°F, but can go as high as 500°F. The higher temperature will cook the vegetables more quickly and produce more charring. It's best to use the lower heat for large vegetables so they have time to cook through.

EXTRA VIRGIN OLIVE OIL Use 1 tablespoon olive oil per cup of vegetables.

PREPARATION Leave the vegetables whole, or cut them into any size you want. The smaller the pieces, the faster they will roast. Use a shallow roasting pan—metal is best except for tomatoes—just large enough to hold the vegetables in a single layer. If the vegetables are piled up, the heat cannot circulate around them and they will steam instead of roast; if the pan is too large, it can overheat and burn the vegetables. Drizzle the olive oil over the vegetables, season with salt and pepper, and, if you like, add whole sprigs or chopped fresh herbs such as rosemary or thyme. Toss the ingredients together well so all the vegetables are coated with oil, and then spread out in a single layer.

ROASTING It can take as little as 15 minutes and as much as an hour for vegetables to roast. The time needed depends on a number of factors: the variety of vegetable, whether they are whole or cut into pieces, the size of the pieces, how brown you want them, etc. Vegetables are done when they are tender enough to eat, but can be cooked longer if you want them more browned or crispier. Be sure and stir the vegetables two or three times during roasting to keep them coated in the olive oil. Most vegetables are roasted in about 30 minutes, but begin to test small pieces after 15 minutes.

STORING Store the roasted and cooled vegetables in individual servings in small freezer bags. When you are ready to use them, put them (still frozen) in a small pan over low heat and cook, turning over now and then, until they are thawed.

CALCULATING CALORIES One tablespoon of olive oil has 120 calories. One cup of vegetables has an average of 20 calories.

Roasted Beets

Roasted Beets are delicious as a side dish all by themselves, but are also a great addition to salads such as the Roasted Beet, Walnut, and Gorgonzola Salad on page 117.

1 unpeeled beet (about 5 ounces) scrubbed clean
2 teaspoons extra virgin olive oil
Salt and black pepper

Equipment needed: roasting pan large enough to hold the beet pieces in one layer

Preheat the oven to 425°F.

Trim off the top and bottom of the beet and cut into 1-inch cubes. Arrange in the pan. Drizzle the olive oil over the beets, season with salt and pepper, and toss to coat with the oil. Spread out into a single layer and roast, turning the beets a few times so they cook evenly, for about 30 minutes, until cooked through.

MAKES I SERVING
CALORIES: 145; STARCH: 0; FAT: ⅔ (olive oil); VEGETABLES: 2 (beets); FRUIT: 0; DAIRY: 0

GOOD TO KNOW

Freezing Vegetables

You can freeze vegetables cooked in olive oil—roasted or pan-cooked—so you always have a good supply in serving-size amounts on hand. Cool the vegetable, then spoon ½- or 1-cup portions into small freezer bags. Lay each bag on its side on the counter, gently flatten to remove any air, and seal. Label the bags, including the amount of oil used, and freeze flat. Once they are frozen, the bags can be stored in rows or stacked. When you want to use the vegetables, do not thaw first. Use scissors to cut down the side seam of the bag and put the frozen block into a small pan and heat over medium-low heat until they are defrosted and warm. If you defrost the vegetables in the bag, the olive oil will stick to the sides of the bag and you will not easily remove it.

Roasted Butternut Squash

If you only know winter squash as the boiled and mashed side dish served at Thanksgiving, give this a try. Roasting concentrates the sweetness and enhances the flavor. The same method can be used for any variety of winter squash.

4 cups ¾-inch-cubed peeled butternut squash (1 medium; about 1½
    pounds)
1 cup 1-inch chunks red onion
3 sprigs fresh thyme
2 tablespoons extra virgin olive oil
Salt and black pepper
Grated nutmeg

Equipment needed: metal roasting pan

Preheat the oven to 450°F.

Combine the squash, onion, and thyme in the pan and drizzle with the olive oil. Season with salt, pepper, and nutmeg and toss together well to coat the squash with the oil. Roast, turning the vegetables a few times, for 20 to 25 minutes, until cooked through and slightly browned. Serve as is or mash.

MAKES 4 SERVINGS
CALORIES: 180; STARCH: 0; VEGETABLES: 2½ (1 onion, 1½ squash); FAT: 1 (olive oil);
FRUIT: 0; DAIRY: 0

Roasted Eggplant

Slices of roasted eggplant are great to have on hand for sandwiches, to chop and add to tomato sauce, and to use as a side dish or a first-course antipasto with other roasted vegetable. Eggplants vary greatly in size, so in order to have the correct calorie count, weigh yours at the market before buying. The one I used for this recipe was about 8 inches long and gave me 16 slices, or 4 per serving.

1 medium unpeeled eggplant, washed (about 1 pound)
¼ cup extra virgin olive oil
Salt

Equipment needed: baking sheet

Preheat the oven to 425°F.

Cut the eggplant crosswise into ½-inch-thick slices. Brush both sides of
each slice with the olive oil and season with salt. Arrange the slices on
the sheet in a single layer. Roast, turning the slices after 15 minutes, for
about 30 minutes, until browned on each side.

MAKES 4 SERVINGS
CALORIES: 150; STARCH: 0; VEGETABLES: 2 (eggplant); FAT: 1 (olive oil); FRUIT: 0; DAIRY: 0

Roasted Green Beans

*Bored with green beans? Roast them. They become a whole different vegetable. In this
recipe, the beans are finished with balsamic vinegar, but you can also use 2 tablespoons
of reduced-sodium soy sauce or a scant 1 tablespoon of lemon juice with a few teaspoons of lemon zest.*

1 cup green beans, washed and trimmed (about 8 ounces)
2 medium garlic cloves, smashed and peeled
1 sprig fresh thyme (optional)
1 tablespoon extra virgin olive oil
Salt and black pepper
3 tablespoons balsamic vinegar

Equipment needed: metal roasting pan that can also go on top of the stove

Preheat the oven to 450°F.

Combine the beans, garlic, and thyme (if using) in the pan. Drizzle with
the olive oil, season with salt and pepper, and toss together until well-

coated with oil. Spread out in one layer and roast for 12 to 15 minutes, until tender and slightly browned.

Remove the pan from the oven and place on the stove over high heat. Add the vinegar and cook, scraping the pan, until the vinegar is almost completely evaporated. Toss and serve.

MAKES I SERVING

CALORIES: 180; STARCH: 0; VEGETABLES: 2 (green beans); FAT: I (olive oil); FRUIT: 0; DAIRY: 0

Roasted Mushrooms

Use these roasted mushrooms as a side dish or scatter them on top of other finished dishes to add a little something extra. Any mushroom can be roasted as below—porcini, cremini, shiitakes, chanterelles, white button. If you slice the mushrooms rather than quarter them, they will become crispier with roasting. Think of them as mushroom chips!

2 cups mushrooms, wiped clean and stems trimmed (about 8 ounces)
2 medium garlic cloves, smashed and peeled
1 tablespoon extra virgin olive oil
Salt and black pepper
1 tablespoon chopped fresh flat-leaf parsley (optional)

Equipment needed: metal roasting pan

Preheat the oven to 450°F.

Cut large and medium mushrooms into quarters; very small mushrooms can be left whole. Combine the mushrooms and garlic in the pan and drizzle with the oil. Season with salt and pepper and toss until all the mushrooms are coated with oil. Spread out in a single layer and roast, turning over a few times, for 10 to 15 minutes, until they are cooked through and lightly browned. Whole mushrooms, depending on the variety, can take up to 30 minutes. Sprinkle with parsley if using and serve.

MAKES I SERVING

CALORIES: 170; STARCH: 0; VEGETABLES: 2 (mushrooms); FAT: I (olive oil); FRUIT: 0; DAIRY: 0

Roasted Peppers

Roasted red and green peppers are practically a "gimme" in a diet. Although they are rubbed with olive oil for roasting, if you peel the skins away after they are cooked, the fat is negligible and need not be counted. Roasted red peppers are interchangeable with "jarred red pepper" in all the recipes. In order to char the skins for easy peeling, roast the peppers at high, 500°F. If you plan to eat the skins, roast in a pan at a lower temperature, 425°F, and only long enough for the peppers to be as tender as you want.

6 bell peppers, red, green, or a combination
2 tablespoons extra virgin olive oil

Preheat the oven to 500°F.

Rub the peppers all over with the olive oil. Place directly on the oven rack and roast, turning often, for 15 to 20 minutes, until the peppers are soft and their skins are shriveled and black. (Use tongs to turn the peppers, and don't squeeze them hard or they will break and the juices will seep out all over your oven.)

Enclose the peppers in a paper bag or place in a bowl and cover with plastic wrap. Let them sit and steam for at least 10 minutes, until they are cool enough to handle and the skins can slip off easily.

Pull the skins off with your fingers and cut the peppers in half lengthwise. Remove the stems, ribs, and seeds. Cut into 1½- or 2-inch-long strips. Store in ⅓ to ½-cup portions. Quantity will depend on how much moisture was lost during roasting. They will keep for up to a week in the refrigerator and 3 months in the freezer.

MAKES 12 SERVINGS, ⅓ TO ½ CUP EACH
CALORIES: 10; STARCH: 0; FAT: 0; VEGETABLES: 1 (pepper); FRUIT: 0; DAIRY: 0

Roasted Potatoes

Remember, potatoes are counted as a starch, not a vegetable. Nevertheless, they are a good side dish. Just be sure to also include vegetables if you have potatoes at lunch or dinner. Roasted potatoes are also a tasty addition to breakfast—a nice starch alternative to toast next to an egg. Want them to taste like home fries? Roast the potatoes with one small onion (peeled and cut into eighths) and toss with a little paprika as well as the salt and pepper.

You can roast any type of potatoes—Red Bliss and Yukon Golds are good choices when you want large pieces; just cut them in halves or quarters. Sprigs of rosemary, thyme, or marjoram add a nice flavor. If you want to make extra to freeze, increase the recipe as you like, keeping the proportion of oil to potatoes at 1 tablespoon oil to 6 ounces potatoes.

6 ounces unpeeled potato, any type, scrubbed clean
1 tablespoon extra virgin olive oil
Salt and black pepper

Equipment needed: metal roasting pan large enough to hold the potatoes in one layer

Preheat the oven to 425°F.

Chop the potato into approximately 1-inch pieces and place in the pan. Drizzle with the olive oil, season with salt and pepper, and toss to coat. Spread out into a single layer and roast for about 30 minutes, until they are cooked through and as brown as you like. The potatoes should be turned periodically while cooking so they evenly brown.

MAKES 1 SERVING
CALORIES: 240; STARCH: 2 (potatoes); FAT: 1 (olive oil); VEGETABLES: 0; FRUIT: 0; DAIRY: 0

Roasted Canned Tomatoes

These roasted tomatoes are delicious as a side dish but also make a great sauce for pasta. I like to use Italian-style tomatoes, which are already diced and seasoned. If you use whole tomatoes, dice them before roasting. You can make them "Italian-style" by roasting two or three along with 2 or 3 peeled garlic cloves and some fresh or dried oregano.

2 (14½-ounce) cans diced plain or Italian-style tomatoes with juices
¼ cup extra virgin olive oil
Salt and black pepper

Equipment needed: 9 × 13-inch non-metal baking pan

Preheat the oven to 425°F.

Pour the tomatoes into the pan. Add the olive oil, season with salt and pepper, and toss together. Roast, stirring occasionally, for 55 to 60 minutes, until most of the liquid is gone and the tomatoes are somewhat shriveled.

MAKES 4 SERVINGS, ABOUT ¾ CUP EACH
CALORIES: 185; STARCH: 0; VEGETABLES: 1½ (tomatoes); FAT: 1 (olive oil); FRUIT: 0; DAIRY: 0

GOOD TO KNOW

The Right Baking Dish

The "right" dish, just like the right tool, makes getting the job done a whole lot easier. I have found that an assortment of small, shallow glass baking dishes with fitted covers is just the right thing for this diet. The ones I have are Pyrex and can go in the oven, the microwave, the freezer, the dishwasher, and to lunch with me because they have very secure fitted covers. Some are rectangular and hold 3 cups; others are round and hold 2 cups. In recipes that call for a small baking dish, this is what I am referring to. After baking, let the dish cool, then cover and either freeze or refrigerate overnight; take along to work the next day to microwave.

White Bean and Tomato Appetizer

This salad is low enough in calories that you can add it as an appetizer to a meal. If you want it as a meal—especially nice in the summer—double all the ingredients.

1 cup mixed salad greens or arugula
¼ cup cooked cannellini beans (page 128), or canned beans, drained and
    rinsed
½ cup chopped fresh tomato (plum, grape, etc.)
3 large olives, pitted and chopped
2 tablespoons finely chopped fresh flat-leaf parsley
1 tablespoon shredded fresh basil
2 teaspoons extra virgin olive oil
Salt and black pepper

Put the greens on a plate. Arrange the beans, tomato, and olives on top. Sprinkle the parsley and basil evenly over the top. Drizzle on the olive oil and season with salt and pepper. Serve at room temperature.

MAKES I SERVING
CALORIES: 175; STARCH: ½ (beans); VEGETABLES: 2 (I each tomato and greens);
FAT: about I (¼ olives, ⅔ olive oil); FRUIT: 0; DAIRY: 0

Cauliflower au Gratin

This gives you a lot of vegetables and not much starch, so you can use it as a filling for a baked potato. You could also enjoy half the recipe as a side dish.

2 tablespoons extra virgin olive oil
½ cup chopped red onion (½ small)
Salt
2 cups cauliflower florets (¼ medium head) or thawed frozen cauliflower
2 tablespoons whole wheat bread crumbs (½ ounce)
¼ cup shredded Cheddar cheese (1 ounce)

Equipment needed: small (5 × 7-inch) baking dish

Heat the olive oil in a medium pan over medium heat. Add the onion, season with salt, and cook, stirring occasionally, until softened but not brown, about 7 minutes. Stir in the cauliflower and cook until the cauliflower is tender and most of the oil has been absorbed, 20 to 25 minutes.

Preheat the broiler. Scrape the vegetables with any oil left in the pan into the baking dish and sprinkle the bread crumbs and cheese on top Place under the broiler and cook until the cheese melts, 5 to 7 minutes.

MAKES I SERVING
CALORIES: 470; STARCH: ½ (bread crumbs); VEGETABLES: 5 (I onion, 4 cauliflower); FAT: 2 (olive oil); FRUIT: 0; DAIRY: I (cheese)

Eggplant "Parmesan"

There is no Parmesan in this recipe but you could substitute some for the mozzarella. Make a large batch and freeze the extra portions; it keeps and reheats beautifully. It is also delicious over pasta.

¾ cup canned diced tomatoes
2 tablespoons extra virgin olive oil
Salt and black pepper
2 thick (about ½-inch) or 3 thin (about ¼-inch) slices unpeeled eggplant
¼ cup shredded part-skim mozzarella cheese (1 ounce)
1 to 2 tablespoons shredded fresh basil leaves

Equipment needed: non-metal roasting pan and small casserole dish

Preheat the oven to 425°F.

Put the tomatoes in the pan and stir in 1 tablespoon of the olive oil. Season with salt and pepper and roast, stirring occasionally, for 30 minutes.

Brush the remaining 1 tablespoon olive oil on both sides of each eggplant slice. Season with salt. After the tomatoes have roasted for 30 minutes, push them to the side of the pan and put the eggplant in the empty spot. Roast for 15 minutes, then carefully flip the eggplant slices over with a spatula and roast for 15 minutes longer, until most of the liquid from the tomatoes has evaporated and the eggplant is browned.

Reduce the temperature to 350°F. Place the eggplant slices in the casserole, overlapping slightly. Sprinkle with the mozzarella, scatter on the basil, and then cover with the tomatoes. Bake for 15 to 20 minutes, until the cheese is melted and the sauce is bubbly.

MAKES 1 SERVING
CALORIES: 350; STARCH: 0; VEGETABLES: 3 (1½ each tomatoes and eggplant);
FAT: 2 (olive oil); FRUIT: 0; DAIRY: 1 (mozzarella)

Grilled Eggplant Dip

Having a party? No need to throw your health to the wind. This dip served with Pita Chips (page 104) is a healthy party food. The instructions below are for grilling the eggplant, but if you prefer, you can roast it instead, following the directions for Roasted Eggplant on page 217. This is a nice time to use your best olive oil.

6 tablespoons extra virgin olive oil
2 (½-inch-thick) slices unpeeled eggplant, 3 to 4 inches in diameter
Salt
1 garlic clove, minced (by hand or through a garlic press)
1 tablespoon chopped fresh thyme
1 cup cooked chickpeas (page 128), or canned chickpeas, drained and
    rinsed
1 tablespoon lemon juice

Preheat the grill to 425°F.

Use 2 tablespoons of the olive oil to brush both sides of each slice of eggplant. Season with salt. Place on the grill and cook, turning once, until both sides are browned, about 5 minutes per side.

Heat 1 tablespoon of the olive oil in a small pan over low heat. Add the garlic and thyme and cook just to flavor the oil, about 3 minutes. Remove from the heat.

Combine the eggplant, garlic, thyme, and oil from the pan with the chickpeas, lemon juice, and ½ teaspoon salt in a food processor fitted with a steel blade. Process until smooth, 3 to 4 minutes. With the machine running, pour the remaining 3 tablespoons oil through the feed tube and continue to process until the oil is incorporated. Serve at room temperature or chilled.

MAKES 4 SERVINGS, ABOUT ½ CUP EACH
CALORIES: 240; STARCH: ½ (beans); VEGETABLES: 1 (eggplant); FAT: 1½ (olive oil); FRUIT: less than ¼ (lemon juice); DAIRY: 0

Mashed Potatoes

If you just have to have totally white mashed potatoes, you can peel them for this recipe. Use any potatoes you like—and, remember, count them as a starch, not a vegetable.

6 ounces potato (1 small baking potato or 2 to 3 red potatoes, depending
    on the size)
Salt
2 tablespoons nonfat or 1% milk, warmed
1 tablespoon extra virgin olive oil

Cut the potato into small (1-inch) pieces. Place in a small saucepan and cover with cold water by 1 inch. Add about ½ teaspoon salt, cover, and bring to a boil. Reduce the heat to medium-low and simmer until the potatoes can be easily pierced with a fork, 5 to 7 minutes.

Drain the potatoes and return to the saucepan. Immediately add the milk and olive oil and mash until the potatoes are the consistency you like.

MAKES 1 SERVING
CALORIES: 250; STARCH: 2 (potatoes); VEGETABLES: 0; FAT: 1 (olive oil); FRUIT: 0; DAIRY: less than ½

Pumpkin with Red Onion

If you are wondering how to use the leftover canned pumpkin from making Pumpkin Bread (page 66), this is it. A nice simple side dish that you can sprinkle with cinnamon or nutmeg.

1 tablespoon extra virgin olive oil
½ cup paper-thin slices red onion (½ small)
¾ cup canned pumpkin

Heat the olive oil in a small pan over medium heat. Add the onion, season with salt, and cook, stirring occasionally, until softened but not brown, about 7 minutes. Stir in the pumpkin and cook for 4 to 5 minutes to heat through.

MAKES I SERVING
CALORIES: 200; STARCH: 0; VEGETABLES: 2½ (1 onion, 1½ pumpkin); FAT: 1 (olive oil);
FRUIT: 0; DAIRY: 0

Vegetable Bread Pudding

This makes 6 servings, so invite five friends over or serve it to your family for brunch. The finished dish does not freeze well but it will keep in the refrigerator for several days. I like it best when it is made with fresh, not frozen, vegetables.

2 cups 1% milk
2 or 3 bay leaves
6 tablespoons plus 2 teaspoons extra virgin olive oil
3 cups chopped fresh broccoli florets (½ small head)
3 cups chopped fresh cauliflower florets (½ small head)
2 cups chopped red onion (1 medium)
Salt and black pepper
6 slices whole wheat bread, slightly stale
6 large eggs, beaten
1½ cups shredded Cheddar cheese (6 ounces)

Equipment needed: 13 × 9-inch baking dish

Combine the milk and bay leaves in a small pan. Heat over low heat until small bubbles appear around the edge of the pan. Remove from the heat and set aside.

Heat the 6 tablespoons of olive oil in a large pan over medium heat. Stir in the broccoli, cauliflower, and onion, season with salt and pepper, and cook until softened, 20 to 25 minutes.

Preheat the oven to 350°F. Brush the 2 teaspoons of olive oil on the bottom and sides of the baking dish.

Layer the bread on the bottom of the pan, overlapping the slices slightly. Remove the bay leaves from the milk. Combine the milk and the eggs in a bowl and stir in the vegetables. Pour the mixture over the bread and sprinkle the cheese over the top. Bake for 35 to 40 minutes, until the center is set. Let stand 5 minutes before cutting.

MAKES 6 SERVINGS
CALORIES: 470; STARCH: 1 (bread); VEGETABLES: 2 ⅔ (⅔ onion, 1 each broccoli and cauliflower); FAT: 1 (olive oil); FRUIT: 0; DAIRY: 1 ⅓ (1 cheese, ⅓ milk)

Yogurt Cucumber Dip

The Greeks call this tzatziki *and serve it as a dip or a sauce as well as a side dish. Put a dollop next to a serving of Zucchini Pancakes (page 229). Mint is traditional and you can use it instead of the dill. The yogurt and cucumbers need to drain before they are used, so plan ahead.*

1 cup plain 1% yogurt
1 cucumber, about 6 inches long (8 ounces), peeled
2 garlic cloves, minced (by hand or through a garlic press)
¼ cup extra virgin olive oil
1 tablespoon lemon juice
1 tablespoon finely chopped fresh dill

Line a colander or strainer with cheesecloth. Spoon in the yogurt and set the colander over a bowl. Let sit for about 2 hours at room temperature, or until most of liquid has drained. The yogurt will appear thicker.

Grate the cucumber or chop into very small pieces. Put in another colander or strainer (without lining) and set in a bowl or the sink. Gently squeeze out as much water as possible. Leave to drain for at least 30 minutes.

Combine the yogurt, cucumber, and garlic in a bowl. Add the olive oil by tablespoons, stirring well with a fork until the oil is thoroughly mixed in before adding more. Stir in the lemon juice and dill. Serve at room temperature or chilled.

MAKES 4 SERVINGS, ABOUT ½ CUP EACH
CALORIES: 160; STARCH: 0; VEGETABLES: 1 (cucumber); FAT: 1 (olive oil);
FRUIT: less than ¼ (lemon juice); DAIRY: ⅛ (yogurt)

Zucchini Pancakes

I couldn't wait to get into my own kitchen and replicate these pancakes after I had them in Greece. They are surprisingly easy to make. Three of the pancakes make a fine appetizer; six of them with Yogurt Cucumber Dip (page 228) and a small salad are a most satisfying lunch.

1 pound zucchini (2 medium)
Salt and black pepper
2 large eggs, beaten
½ cup finely chopped fresh flat-leaf parsley
½ cup finely chopped red onion (½ small)
½ cup whole wheat bread crumbs
½ cup shredded Cheddar cheese (2 ounces)
1 cup extra virgin olive oil

Equipment needed: baking sheet

Shred the zucchini by hand or with a food processor fitted with a shredding blade. You will have about 4 cups. Put the zucchini in a colander set in a bowl or the sink and sprinkle with salt. Use a fork to toss and distribute the salt. Let drain for 2 hours.

Combine the drained zucchini, eggs, parsley, onion, bread crumbs, and Cheddar in a bowl and stir to combine. Season with salt and pepper.

Cover a baking sheet with paper towels and keep near the stove. Heat the olive oil in a deep skillet over medium heat until hot. It will bubble slightly and feel warm when you hold your hand an inch above the pan. Drop tablespoons of the batter into the hot oil. Cook until slightly brown, 3 to 4 minutes. Turn the pancakes over using a spatula and cook the other sides until browned, 3 to 4 minutes. Transfer to the paper towels. Wait a few minutes before adding more batter so the oil can return to the correct temperature.

Let the pancakes cool slightly on the paper towels before serving.

MAKES 5 SERVINGS, 6 PANCAKES EACH
CALORIES: 300; STARCH: less than ½; VEGETABLES: 2 (⅕ parsley, ⅕ onion, 1 ⅔ zucchini); FAT: 1 ⅔ (olive oil); FRUIT: 0; DAIRY: less than ½

11

Desserts

You can definitely eat dessert while you are on this diet—just not the wrong desserts. What you buy in a bakery or at the grocery store contains, more times than not, what you shouldn't be eating and is swathed in more calories than you can afford.

All the desserts in this chapter are made with ingredients that are beneficial to you. None of them have more than 300 calories, a manageable number that fits into your meal plan. Some of your daily food plans will have fewer than the allowed 1,500 calories; the extra calories are there for optional food choices and a dessert can be one of those options. On other days, you may have to eliminate other foods to allow for the sweet. I recommend that you adjust the amount of starch and/or fruit for that day. It is best to make the change on the day you eat the dessert, but if it doesn't happen that way and you wind up eating 1,800 calories in a day, eat less the next day.

You can modify your own dessert recipes to make them healthier. I do it all the time. It may involve a bit of trial and error, but here are some tips that will get you started:

- You can substitute extra virgin olive oil for any fat called for in a recipe. So out with the vegetable oil and margarine and in with nutrient-rich olive oil. Use a mild flavored olive oil, and it need not be your best one.
- In cookies and bars, whole wheat flour can actually improve taste and texture, so substitute at least half of the all-purpose flour with whole wheat flour. Cakes are more difficult: You can use some whole wheat pastry flour, but too high a ratio makes a denser product, which is not what most cakes should be—

unless, of course, it is a fruit cake for which "dense" is the description.

- Substitute brown sugar for at least some, if not all, of the white granulated sugar in the recipe. Brown sugar is merely white sugar with molasses and is not, as some people think, a "healthy food." But brown sugar does provide extra moisture, which is needed when whole wheat is present.

- If you have a number of favorite dessert recipes that you would like to convert, I recommend *Olive Oil Baking* by Lisa A. Sheldon. It includes useful tips and techniques for converting recipes, as well as a number of original recipes for popular desserts.

Cranberry and Almond Biscotti

Three of these make a satisfying, quick, and healthy breakfast. For holiday color, use pistachios in place of the almonds. No need to toast them (as the almonds below); just add them with the cranberries. With pistachios, there are 95 calories per bar.

1½ cups whole almonds with skins
1½ cups all-purpose flour
1 cup whole wheat flour
¾ cup granulated sugar
¾ cup loosely packed brown sugar
½ cup extra virgin olive oil
2 large eggs, beaten
1 tablespoon water
4 teaspoons baking powder
1 teaspoon almond extract
1 teaspoon cinnamon
½ teaspoon ground cloves
1 cup dried cranberries

Equipment needed: baking pan and 2 nonstick baking sheets, 14 × 16½ inches

Preheat the oven to 300°F. Spread the almonds in a single layer in the baking pan and bake for 5 minutes, or until lightly browned. Keep close watch because nuts can burn very quickly. Transfer the pan to a wire rack and let the almonds cool. Increase the oven temperature to 375°F.

With an electric mixer on medium speed, mix the all-purpose flour, whole wheat flour, granulated sugar, brown sugar, olive oil, eggs, water, baking powder, almond extract, cinnamon, and cloves until homogeneous, 3 to 4 minutes. The dough will be sticky. Add the almonds and cranberries and mix on low until evenly distributed.

Divide the dough into 4 equal pieces. On each of the 2 baking sheets, shape 2 portions into logs about 14 inches long and 2 inches wide, keeping them well apart from each other.

Place the cookie sheets on separate oven racks and position them off to each side so they are not directly above or below each other. Bake for 12 minutes, then swap the sheets from one position to the other. Continue baking for 10 to 12 minutes longer, until the logs are firm to the touch.

Transfer the sheets to wire racks. Use a metal spatula to gently loosen the bottoms of the logs, but leave them on the sheets. Let cool for 10 minutes.

Cut each log crosswise into 12 bars and nudge apart so there is space between them. Return the biscotti to the oven, again staggering the sheet positions, and bake for 5 minutes. Swap the sheets and bake for 5 minutes longer, until the pieces are lightly browned. Cool on wire racks, then store in a covered container.

MAKES 48 BARS
CALORIES: 100; STARCH: less than 1; VEGETABLES: 0; FAT: slightly more than ½
(½ nuts, some olive oil); FRUIT: less than 1; DAIRY: 0

Chocolate Biscotti

The name of this traditional Italian treat means "twice baked"; rightly so, as the cookies need to be baked twice in order to get properly crunchy. For the cocoa powder, use one that is marked "dark"; it will give a richer flavor. Do not use cocoa mix.

1½ cups all-purpose flour
1 cup whole wheat flour
¾ cup granulated sugar
¾ cup loosely packed brown sugar
¾ cup extra virgin olive oil
½ cup cocoa powder
3 large eggs, beaten
1 tablespoon water
4 teaspoons baking powder
1 teaspoon almond extract

Equipment needed: 2 nonstick baking sheets, 14 × 16½ inches

Preheat the oven to 375°F.

Using an electric beater on medium speed, blend all the ingredients together until well combined, 3 to 4 minutes. The dough will be sticky.

Divide the dough into 4 equal pieces. On each of the 2 baking sheets, shape 2 portions into logs about 14 inches long and 2 inches wide, keeping them well apart from each other.

Place the baking sheets on separate oven racks and position them off to each side so they are not directly above or below each other. Bake for 12 minutes, then swap the sheets from one position to the other and continue baking for 12 minutes longer, or until the logs are firm to the touch.

Transfer the cookie sheets to wire racks. Use a metal spatula to gently free the bottoms of the logs, but leave them on the sheets. Let cool for 10 minutes.

Cut each log crosswise into 12 bars and nudge the bars apart so there is space between them and the heat of the oven can reach the sides. Return the biscotti to the oven, again staggering the sheet positions, and bake for 5 minutes. Swap the sheets and bake 5 minutes longer, until slightly toasted.

Transfer the biscotti to wire racks to cool. Store in a covered container.

MAKES 48 BARS
CALORIES: 80; STARCH: less than 1; VEGETABLES: 0; FAT: less than 1; FRUIT: less than 1; DAIRY: 0

Chocolate Cake

This cake gets its moistness from zucchini the same way carrot cake gets it from carrots. Dark cocoa is best, but if you can't find it, regular cocoa (not the hot drink mix) is fine. A dollop of plain unsweetened yogurt and some fresh berries make a nice topping on each piece (remember to count the dairy and the fruit).

DRY INGREDIENTS
1 cup whole wheat flour
1 cup all-purpose flour
½ cup cocoa powder
2 teaspoons baking powder
1 teaspoon salt
1 cup chopped walnuts

LIQUID INGREDIENTS
1 large egg
2 cups shredded unpeeled zucchini (about 8 ounces)
1½ cups loosely packed brown sugar
1 cup extra virgin olive oil
¼ cup plain low-fat yogurt
2 teaspoons vanilla extract
1 teaspoon almond extract

Equipment needed: 9-inch square nonstick cake pan

Preheat the oven to 350°F.

Combine the dry ingredients, except for the walnuts, in a large bowl. Use a fork to thoroughly mix, then stir in the nuts. Whisk the egg in another bowl until liquid, then add the remaining liquid ingredients. Whisk until completely combined.

Make a well in the center of the dry ingredients and pour in the liquid ingredients. Use a rubber spatula to mix the ingredients, gently, just until combined.

Pour the batter into the cake pan and bake for 30 minutes, or until the top springs back when touched. Cool before cutting into 16 equal squares.

MAKES 16 PIECES
CALORIES: 290; STARCH: about 1; VEGETABLES: less than 1; FAT: 1½ (½ nuts, 1 olive oil); FRUIT: 0; DAIRY: 0

Hermits

I love hermits. My mother always made them when I was a kid. If you are unfamiliar with the spicy, molasses-flavored bars ubiquitous in New England, now is a good time to discover them. They make a nice breakfast on the run.

I changed my mother's recipe to use olive oil and whole wheat flour, and now I love them even more knowing how nutritious they are. Note that the batter is quite thick and a standing mixer handles the job well. It is possible to mix with a handheld mixer, but your arm does get tired.

1½ cups whole wheat flour
1½ cups all-purpose flour
½ teaspoon nutmeg
½ teaspoon cinnamon
½ teaspoon ground cloves
½ teaspoon salt
½ cup extra virgin olive oil
1 cup loosely packed brown sugar

½ cup 1% milk
½ cup molasses
1 cup raisins
½ cup chopped walnuts

Equipment needed: standing mixer and 2 nonstick baking sheets, 14 × 16½ inches

Preheat the oven to 350°F.

Thoroughly combine the whole wheat flour, all-purpose flour, nutmeg, cinnamon, cloves, and salt in a large bowl with a fork or whisk.

Use a mixer fitted with a paddle on medium speed to thoroughly combine the olive oil and brown sugar, 3 to 4 minutes. The mixture will be thick but smooth. With the mixer on low speed, gradually pour in the milk and molasses, then increase the speed to medium and beat until thoroughly combined, 3 to 4 minutes. Turn the mixer back to low speed and gradually add the dry ingredients about ½ cup at a time. When all the dry ingredients have been added, increase the speed to medium and beat until the batter is homogeneous, about 3 minutes. Remove the mixing bowl from the stand and stir in the raisins and nuts with a rubber spatula.

Divide the dough into 4 portions; place 2 portions on each baking sheet. Spread each piece the length of the cookie sheet and shape into a flattened log about 14 inches long, 3 inches wide, and 1 inch high. Use a rubber spatula or your hands to mold and smooth out the logs.

Bake for 18 to 22 minutes, until the dough is set but still slightly soft to touch. Transfer the sheets to wire racks and cool the logs on the sheets for 5 minutes. Cut each log into 9 bars, each about 1½ inches thick.

MAKES 36 BARS

CALORIES: 114; STARCH: less than 1; VEGETABLES: 0; FAT: less than 1; FRUIT: less than 1; DAIRY: 0

Pumpkin Pie

Delicious and nutrient-rich, this pie is as good for breakfast as it is for dessert. For the pumpkin, you want a can of just pumpkin—no seasonings or sweetener.

CRUST

1 cup dry old-fashioned (rolled) oats
½ cup loosely packed brown sugar
6 tablespoons extra virgin olive oil

FILLING

2 large eggs
1¾ cups canned pumpkin (15-ounce can)
1½ cups 1% milk
⅔ cup loosely packed brown sugar
1 teaspoon cinnamon
½ teaspoon ground ginger
½ teaspoon nutmeg
½ teaspoon salt

Equipment needed: 9-inch pie pan

Preheat the oven to 425°F.

To prepare the crust: Use a fork to mix together the oats, brown sugar, and olive oil in a bowl. Gently press the mixture into the pie pan using your fingers to make an even layer that goes up the side of the pan.

To make the filling: Beat the eggs on medium speed with an electric mixer until liquid, 2 to 3 minutes. Add the pumpkin, milk, brown sugar, cinnamon, ginger, nutmeg, and salt and mix until smooth, 3 to 4 minutes.

Pour the filling into the crust and bake for 15 minutes. Reduce the heat to 350°F and bake for 30 to 35 minutes longer, until the center of the pie is set.

Cool the pie on a rack about 5 minutes, then run a sharp knife along the side of the pan to gently separate the crust from the pan. This will make it easier to cut when cool.

MAKES 8 SERVINGS

CALORIES: 260; STARCH: ½ (oats); VEGETABLES: less than 1; FAT: ¾ (olive oil); FRUIT: 0; DAIRY: less than 1

GOOD TO KNOW

Green Tea

After water, tea is the most consumed drink in the world. Green tea has been part of Chinese medicine since ancient times—and probably with good reason. It contains catechins, a type of phytonutrient that is a strong antioxidant. Catechins are found predominantly in green tea; smaller amounts are also in black tea, grapes, wine, and chocolate.

Green tea research suggests that it has properties that are protective against cancer—note that "suggests" does not mean "proves." More research needs to be done, but individual studies have found that green tea can lessen cell damage from oxidation, inhibit tumor growth in animals, and slightly decrease breast cancer risk. So should you drink it? Three or four cups of green tea a day are not going to hurt you and may well prove to be cancer protective. The time-out moments you spend drinking tea can't hurt either.

Glossary

Antioxidants: Antioxidants are compounds that can stop the damage caused by free radicals, which are molecules in our bodies that have one or more unpaired electrons, caused when food is metabolized. Molecules are uncomfortable with unpaired electrons, so they grab electrons from things they are near: cell membranes or DNA. The place where the electrons are taken from is damaged—cell membranes that have lost electrons are not as sturdy; damaged DNA can contribute to cancer. Antioxidants to the rescue. They donate electrons to the free radicals, thereby saving the cell membrane or DNA from harm. This is why it is so important to include foods high in antioxidants—such as extra virgin olive oil—in your diet.

Biomarkers: Biomarkers are measureable indicators in our blood and urine. Determining the levels of biomarkers is the objective of clinical tests such as blood work and urine samples. Certain biomarkers are related to particular chronic disease development and could be called risk factors for that disease. As a general rule, the more biomarkers you have for a disease, the greater your chances of getting that disease. Some biomarkers for breast cancer are insulin resistance, higher levels of fasting blood glucose, and an above-healthy level of inflammation.

Carotenoid: Carotenoids are compounds that give pigment or color to plant products; the deeper the color, the higher the amount of carotenoid. The colors can be red, orange, or yellow. Many vegetables that are deep green (such as broccoli, spinach, and kale) also have high levels of carotenoids, but the green is from chlorophyll (the part of the plant needed for photosynthesis). Many studies have shown that carotenoids have a number of health benefits, especially for fighting cancer. Those studies, however, measure the blood levels of carotenoids, not the dietary intake of vegetables and fruits that contain them. Eating lots of fruits and vegetables will not alone raise your carotenoid level.

Carotenoids are digested and absorbed with fat, so you need to eat fat with the carotenoid to get it into your body.

Cruciferous vegetables: Cruciferous vegetables, *Brassica oleracea,* are in the cabbage family. The name *cruciferous* means cross-bearing and refers to the four-petal flowers that form a cross. Some of the vegetables in this family are broccoli, Brussels sprouts, cabbage, cauliflower, kale, kohlrabi, and Swiss chard. All contain a sulfur-containing phytonutrient family called the glucosinolates. When cruciferous vegetables are chopped or chewed, glucosinolates break down (hydrolysis), producing a phytonutrient called isothiocyanate. Studies have shown that isothiocyanates have powerful cancer-fighting properties, including the ability to prevent cancers from starting and spreading. Isothiocyanate seems to be especially effective in fighting breast and prostate cancers.

Estrogen: Estrogen is a hormone that has a good news/bad news story. Estrogen is necessary for the development of female characteristics, but a high lifetime exposure to estrogen's actions has been shown to increase breast cancer risk. Estrogen is usually metabolized to forms that have different functions, some of which can increase breast cancer risk. The cruciferous vegetables contain a family of phytonutrients called the glucosinolates, which work to produce the form of estrogen that is not related to breast cancer.

Glucosinolates: These are a phytonutrient family found in the *Brassica,* or cabbage, family of vegetables (broccoli, Brussels sprouts, cauliflower, kohlrabi, kale, turnip). Glucosinolates contains sulfur, which is responsible for the bitter taste of these vegetables. Glucosinolates are hydrolyzed in the body to make compounds (isothiocyanates) that have been shown to stop cancer from spreading and to activate enzymes that halt the making of carcinogens. Glucosinolates also alter estrogen metabolism by encouraging the production of 2-hydroxyestrone, a form of estrogen that does not promote breast cancer. Because glucosinolates are water soluble, cooking the vegetable containing them in olive oil ensures that they get into your body.

Insulin: Insulin is a hormone made by the pancreas and released into the blood mainly when there is a rise in blood glucose (a form of carbohydrate also called blood sugar), such as after a meal. Insulin's job is to decrease blood levels of nutrients by ushering them into the cells for storage. Insulin is needed to get glucose into the fat and muscle cells and to get amino acids into muscle cells, where they can be stored as protein.

Insulin resistance: Insulin resistance is the term used when cells no longer completely respond to insulin. In the beginning stages of insulin resistance, the pancreas sends out more insulin and blood glucose levels may increase but not above a healthy level (100 mg/dl for fasting blood glucose). A person can go for several years with insulin resistance but not know it because the blood glucose stays below a level that would raise concern. Eventually, however, the pancreas cannot keep up and the blood glucose levels increase to where the patient is labeled prediabetic, which is when fasting blood glucose is 100 to 125 mg/dl. If the fasting blood glucose is more than 126 mg/dl, the person is now classified as diabetic (type 2). Both extra insulin and glucose increase cancer risk.

Lignans: Lignans are a noncarbohydrate component of fiber. Found in the seeds of fruits (berries), flax, and sesame seed and in the bran part of whole grains, lignans are a class of phytoestrogens that have been related to breast cancer protection.

Lycopene: Lycopene is a type of carotenoid (family of phytonutrients). Lycopene provides red color to plant products. The main dietary source of lycopene is tomatoes, but it is also found in some other red produce, such as pink grapefruit and watermelon (but not cherries or strawberries). A number of studies have shown that lycopene can reduce cancer risk. However, other studies do not, and these results could be due to the food source of lycopene, the way the food is prepared, or both. Lycopene in fresh tomatoes is not well absorbed; the tomatoes need to be canned, cooked, or processed to make lycopene in a form that can be absorbed. Also, like all carotenoids, lycopene needs fat to be absorbed, so unless the food containing lycopene is eaten with fat, only very small amounts will be absorbed.

Monounsaturated fat: Monounsaturated fat is a type of fatty acid with a single double bond, hence the *mono* in the term. Fatty acids are chains of carbons surrounded by hydrogen. Every carbon must make four bonds; if hydrogens are missing from two adjacent carbons, those two carbons form a double bond and you have a polyunsaturated fat, which oxidizes. (*See* oxidation.) Monounsaturated fats do not oxidize—a very good thing. Monounsaturated fats are liquid at room temperature and semisolid or solid when refrigerated.

Oleocanthal: Oleocanthal is a phytonutrient in olive oil that provides the distinctive taste in the oil. Studies have shown that oleocanthal works like nonsteroidal anti-inflammatory drugs (NSAIDS), for example, ibuprophen. NSAIDS decrease inflammation by blocking an enzyme that leads to inflammation.

Oxidation: Oxidation is the reaction in which a molecule loses an electron and becomes desperate to find a new one. When this happens, the cell is said to be oxidized and is called a free radical. Radical indeed, because with only one electron, it grabs electrons from the closest source, such as cell membranes or DNA. The place from which the free radicals take the electrons, however, is damaged. Excessive oxidation contributes to aging (of all cells) and all chronic diseases, including cancer and heart disease. Need a more graphic picture? Think of rust on a bicycle fender, a now green but once shiny copper penny, or a lovely cut fresh apple suddenly turned brown—all because of oxidation. We cannot stop oxidation, but we can minimize it by eating a diet that is low in polyunsaturated fats and high in extra virgin olive oil and food sources of antioxidants.

Polyunsaturated fat: Polyunsaturated fats are a type of fatty acid with two or more double bonds—hence *poly* in the term. Fatty acids are chains of carbons surrounded by hydrogen. Every carbon needs to make four bonds, so if hydrogen is missing from two adjacent carbons, the carbons make a double bond. Polyunsaturated fat oxidizes.

Phytonutrients: Phytonutrients (also called phytochemicals) are active compounds found in all plant products. They are not essential nutrients such as vitamins or minerals because they are not involved in growth or tissue maintenance. Oh, but what they can do! Phytonutrients protect the plant they are in from the environment. Think of all those plants that have survived in spite of atrocious mistreatment by the weather, crawly things, and careless gardeners. (Okay, not all my tomatoes survived the leaf miners.) Because the enemy out to get them is in the environment, the content of phytonutrients is usually higher on the outside of the plant, which is why it is good to eat the skins of fruits and vegetables and the entire grain of wheat products. When we eat food containing phytonutrients, we are ingesting their properties that can protect us. Studies in test tubes have shown that phytonutrients have amazing properties that could work to decrease risk factors or biomarkers of all chronic diseases.

Saturated fat: Saturated fats are a type of fatty acid with no double bonds in the carbon chain. They are found mainly in animal fat (meat, full-fat dairy) but are also in tropical oils (coconut and palm). Saturated fats are solid at room temperature and do not oxidize.

Squalene: Squalene is a compound found in extra virgin olive oil (and shark liver oil). It is one of the compounds our bodies use to make cholesterol and other steroid compounds. Squalene has been related to tumor inhibition, and studies suggest that the squalene content of olive oil accounts in large part for the association of olive oil with diminished cancer risk.

Trans fatty acids: The definition of trans fatty acids (also known as *trans fats*) should begin and end with "Very, very bad. Don't eat them." A high intake of trans fatty acids has been related to all chronic diseases—*all*. Trans fats are created by adding hydrogen to vegetable oil (hydrogenation), thereby making a liquid oil more solid. If trans fats are so bad, why do we have them? Trans fats are less likely to spoil, so foods that contain them have a longer shelf life. In addition, foods that are high in fat *feel* less greasy. Enough said?

Triglycerides: Triglycerides are a type of fat (lipid) in our blood that can be measured. If your triglyceride level is high (more than 100 mg/dl) af-

ter a ten-hour fast, it usually means that your insulin is not working well
or you are becoming insulin resistant. Fasting triglycerides higher than
100 can be a risk factor, or biomarker, for breast cancer. The higher
triglycerides mean not that you will definitely get breast cancer but that
your risk of breast cancer is increased.

Notes

Introduction

1. K. Kerlikowske, R. Walker, D.L. Miglioretti, A. Desai, R. Ballard-Barbash, and D.S. Buist, "Obesity, mammography use and accuracy, and advanced breast cancer risk," *Journal of the National Cancer Institute,* 100 (23), December 3, 2008: 1724–1733.

2. C.H. Kroenke, W.Y. Chen, B. Rosner, and M.D. Holmes, "Weight, weight gain, and survival after breast cancer diagnosis," *Journal of Clinical Oncology,* 23 (7), March 1, 2005: 1370–1378.

3. R.L. Prentice, B. Caan, R.T. Chlebowski, et al., "Low-fat dietary pattern and risk of invasive breast cancer: the Women's Health Initiative Randomized Controlled Dietary Modification Trial," *Jama,* 295 (6), February 8, 2006: 629–642.

4. D.J. Hunter, D. Spiegelman, H.O. Adami, et al., "Cohort studies of fat intake and the risk of breast cancer—a pooled analysis," *New England Journal of Medicine,* 334 (6), February 8, 1996: 356–361.

5. M.S. Donaldson, "Nutrition and cancer: a review of the evidence for an anti-cancer diet," *Nutrition Journal,* 3, October 20, 2004: 19.

6. M.J. Brown, M.G. Ferruzzi, M.L. Nguyen, et al., "Carotenoid bioavailability is higher from salads ingested with full-fat than with fat-reduced salad dressings as measured with electrochemical detection," *American Journal of Clinical Nutrition,* 80 (2), August 2004: 396–403. A.W. Williams, T.W. Boileau, and J.W. Erdman, "Factors influencing the uptake and absorption of carotenoids," *Proceeding of the Society for Experimental Biology and Medicine,* 218 (2), June 1998: 106–108.

7. W.C. Willett, "Is dietary fat a major determinant of body fat?" *American Journal of Clinical Nutrition,* 67 (3 Suppl), March 1998: 556S–562S.

8. K. Vigilante and M. Flynn, *Low-Fat Lies, High-Fat Frauds, and the Healthiest Diet in the World* (Washington, DC: LifeLine Press, 1999).

9. I. Shai, D. Schwarzfuchs, Y. Henkin, et al., "Weight loss with a low-carbohydrate, Mediterranean, or low-fat diet," *New England Journal of Medicine,* 359 (3), July 17, 2008: 229–241.

10. K. Hobson, "Diets That Promote Health (and Always Have)," *US News & World Report* (2008).

11. Ibid.

Chapter 1

1. W. Demark-Wahnefried, B.L. Peterson, E.P. Winer, et al., "Changes in weight, body composition, and factors influencing energy balance among pre-menopausal breast cancer patients receiving adjuvant chemotherapy," *Journal of Clinical Oncology,* 19 (9), May 1, 2001: 2381–2389.

2. M.L. Irwin, A. McTiernan, R.N. Baumgartner, et al., "Changes in body fat and weight after a breast cancer diagnosis: influence of demographic, prognostic, and lifestyle factors," *Journal of Clinical Oncology,* 23 (4), February 1, 2005: 774–782.

3. Breast Cancer Weight Gain, American Cancer Society, www.cancer.org/docroot/NWS/, accessed October 7, 2008.

4. M.L. Irwin, A. McTiernan, R.N. Baumgartner, et al., "Changes in body fat and weight after a breast cancer diagnosis."

Chapter 2

1. R.W. Owen, R. Haubner, G. Wurtele, E. Hull, B. Spiegelhalder, and H. Bartsch. "Olives and olive oil in cancer prevention," *European Journal of Cancer Prevention,* 13 (4), August 2004: 319–326.

2. C. la Vecchia, E. Negri, S. Franceschi, A. Decarli, A. Giacosa, and L. Lipworth, "Olive oil, other dietary fats, and the risk of breast cancer (Italy)" *Cancer Causes Control,* 6 (6), November 1995: 545–550.

3. J.M. Martin-Moreno, W.C. Willett, L. Gorgojo, et al., "Dietary fat, olive oil intake and breast cancer risk," *International Journal of Cancer,* 58 (6), September 15, 1994: 774–780.

4. A. Trichopoulou, K. Katsouyanni, S. Stuver, et al., "Consumption of olive oil and specific food groups in relation to breast cancer risk in Greece," *Journal of the National Cancer Institute,* 87 (2), January 18, 1995: 110–116.

5. R.E. Harris, K.K. Namboodiri, and W.B. Farrar, "Nonsteroidal anti-inflammatory drugs and breast cancer," *Epidemiology,* 7 (2), March 1996: 203205. B. Takkouche, C. Regueira-Mendez, and M. Etminan, "Breast cancer and use of nonsteroidal anti-inflammatory drugs: a meta-analysis," *Journal of the National Cancer Institute,* 100 (20), October 15, 2008: 1439–1447. Y.S. Zhao, S. Zhu, X.W. Li, et al., "Association between NSAIDs use and breast cancer risk: a systematic review and meta-analysis," *Breast Cancer Research and Treatment,* November 2, 2008.

6. M.L. Kwan, L.A. Habel, M. Slattery, and B. Caan, "NSAIDs and breast cancer recurrence in a prospective cohort study," *Cancer Causes Control,* 18 (6), August 2007: 613–620.

7. M.-I. Covas, V. Ruiz-Gutierrez, R. de la Torre, A. Kafatos, R. Lamuela-Raventos, J. Osada, R.W. Owen, F. Visioli, "Minor components of olive oil: evidence to date of health benefits in humans," *Nutrition Reviews,* 64 (10), 2006: S20–S30.

8. N.M. Saarinen, A. Warri, M. Airio, A. Smeds, S. Makela, "Role of dietary lignans in the reduction of breast cancer risk," *Molecular Nutrition & Food Research,* 51 (7), July 2007: 857–866.

9. L.S. Velentzis, M.M. Cantwell, C. Cardwell, M.R. Keshtgar, A.J. Leathem, and J.V. Woodside, "Lignans and breast cancer risk in pre- and post-menopausal women: meta-analyses of observational studies," *British Journal of Cancer,* 100 (9), May 5, 2009: 1492–1498.

10. B.N. Fink, S.E. Steck, M.S. Wolff, et al., "Dietary flavonoid intake and breast cancer risk among women on Long Island," *American Journal of Epidemiology,* 165 (5), March 1, 2007: 514–523.

11. M.J. Gunter, D.R. Hoover, H. Yu, et al., "Insulin, insulin-like growth factor-I, and risk of breast cancer in postmenopausal women," *Journal of the National Cancer Institute,* 101 (1), January 7, 2009: 48–60.

12. J. Sabate, "Nut consumption and body weight," *American Journal of Clinical Nutrition,* 78 (3 Suppl), September 2003: 647S–650S.

13. M. Bes-Rastrollo, J. Sabate, E. Gomez-Gracia, A. Alonso, J.A. Martinez, and M.A. Martinez-Gonzalez, "Nut consumption and weight gain in a Mediterranean cohort: The SUN study," *Obesity (Silver Spring),* 15 (1), January 2007: 107–116.

14. S. Liu, W.C. Willett, J.E. Manson, F.B. Hu, B. Rosner, and G. Colditz, "Relation between changes in intakes of dietary fiber and grain products and changes in weight and development of obesity among middle-aged women," *American Journal of Clinical Nutrition,* 78 (5), November 2003: 920–927.

15. C.K. Good, N. Holschuh, A.M. Albertson, and A.L. Eldridge, "Whole grain consumption and body mass index in adult women: an analysis of NHANES 1999–2000 and the USDA pyramid servings database," *Journal of the American College of Nutrition,* 27 (1), February 2008: 80–87.

16. A. Lee, D.I. Thurnham, and M. Chopra, "Consumption of tomato products with olive oil but not sunflower oil increases the antioxidant activity of plasma," *Free Radical Biology & Medicine,* 29 (10), November 15, 2000: 1051–1055.

17. B. Challier, J.M. Perarnau, and J.F. Viel, "Garlic, onion and cereal fibre as protective factors for breast cancer: a French case-control study," *European Journal of Epidemiology,* 14 (8), December 1998: 737–747.

18. H.A. Bischoff-Ferrari, B. Dawson-Hughes, J.A. Baron, et al., "Calcium intake and hip fracture risk in men and women: a meta-analysis of prospective cohort studies and randomized controlled trials," *American Journal of Clinical Nutrition,* 86 (6), December 2007: 1780–1790.

19. D. Feskanich, W.C. Willett, M.J. Stampfer, G.A. Colditz, "Protein consumption and bone fractures in women," *American Journal of Epidemiology,* 143 (5), March 1, 1996: 472–479.

20. S.A. Missmer, S.A. Smith-Warner, D. Spiegelman, et al., "Meat and dairy food consumption and breast cancer: a pooled analysis of cohort studies," *International Journal of Epidemiology,* 31 (1), February 2002: 78–85. P.G. Moorman and P.D. Terry, "Consumption of dairy products and the risk of breast cancer: a review of the literature," *American Journal of Clinical Nutrition,* 80 (1), July 2004: 5–14. P.W. Parodi, "Dairy product consumption and the risk of breast cancer," *Journal of the American College of Nutrition,* 24 (6 Suppl.), December 2005: 556S–568S.

21. M.H. Shin, M.D. Holmes, S.E. Hankinson, K. Wu, G.A. Colditz, and W.C. Willett. "Intake of dairy products, calcium, and vitamin D and risk of breast cancer," *Journal of the National Cancer Institute,* 94 (17), September 4, 2002: 1301–1311.

22. A.L. Frazier, C.T. Ryan, H. Rockett, W.C. Willett, G.A. Colditz, "Adolescent diet and risk of breast cancer," *Breast Cancer Research,* 5 (2), 2003: R59–64. J. Shannon, R. Ray, C. Wu, et al., "Food and botanical groupings and risk of breast cancer: a case-control study in Shanghai, China," *Cancer Epidemiology, Biomarkers & Prevention,* 14 (1), January 2005: 81–90.

23. M.A. Murtaugh, J.S. Herrick, C. Sweeney, et al., "Diet composition and risk of overweight and obesity in women living in the southwestern United States," *Journal of the American Dietetic Association,* 107 (8), August 2007: 1311–1321.

24. M. Kapiszewska, "A vegetable to meat consumption ratio as a relevant factor determining cancer preventive diet. The Mediterranean versus other European countries," *Forum of Nutrition,* 59, 2006: 130–153.

25. G.C. Kabat, A.J. Cross, Y. Park, et al., "Meat intake and meat preparation in relation to risk of postmenopausal breast cancer in the NIH–AARP diet and health study," *International Journal of Cancer,* 124 (10), May 15, 2009: 2430–2435.

26. K. el-Bayoumy, Y.H. Chae, P. Upadhyaya, et al., "Comparative tumorigenicity of benzo[a]pyrene, 1-nitropyrene and 2-amino-1-methyl-6-phenylimidazo[4,5-b]pyridine administered by gavage to female CD rats," *Carcinogenesis,* 16 (2), February 1995: 431–434. E.G. Snyderwine, M. Venugopal, and M. Yu, "Mammary gland carcinogenesis by food-derived heterocyclic amines and studies on the

mechanisms of carcinogenesis of 2-amino-1-methyl-6-phenylimidazo[4,5-b]pyridine (PhIP)," *Mutation Research,* 506–507, September 30, 2002: 145–152.

27. W. Zheng, D.R. Gustafson, R. Sinha, et al., "Well-done meat intake and the risk of breast cancer," *Journal of the National Cancer Institute,* 90 (22), November 18, 1998: 1724–1729.

28. H. Kuper, W. Ye, E. Weiderpass, et al., "Alcohol and breast cancer risk: the alcoholism paradox," *British Journal of Cancer,* 83 (7), October 2000: 949–951.

29. F. Bessaoud and J.P. Daures, "Patterns of alcohol (especially wine) consumption and breast cancer risk: a case-control study among a population in Southern France," *Annals Epidemiology,* 18 (6), June 2008: 467–475..

30. K.W. Reding, J.R. Daling, D.R. Doody, C.A. O'Brien, P.R. Porter, and K.E. Malone, "Effect of prediagnostic alcohol consumption on survival after breast cancer in young women," *Cancer Epidemiology, Biomarkers & Prevention,* 17 (8), August 2008: 1988–1996.

31. J.O. Hill and H.R. Wyatt, "Role of physical activity in preventing and treating obesity," *Journal of Applied Physiology,* 99 (2), August 2005: 765–770.

Chapter 4

1. Y. Ma, E.R. Bertone, E.J. Stanek III, et al. "Association between eating patterns and obesity in a free-living US adult population," *American Journal of Epidemiology,* 158(1), July 1, 2003:85–92.

Chapter 10

1. M. Pollan, *In Defense of Food* (New York: Penguin Books, 2009).

Bibliography

Barr, Nancy Verde. *We Called It Macaroni.* New York: Knopf, 1990.

_____. *Make It Italian.* New York: Knopf, 2002.

Herbst, Sharon Tyler. *Food Lover's Companion.* 3rd Edition. New York: Barron's, 2001.

Krasner, Deborah. *The Flavors of Olive Oil: A Tasting Guide and Cookbook.* New York: Simon & Schuster, 2002.

Lucas, Geralyn. *Why I Wore Lipstick to My Mastectomy.* New York: St. Martin's Griffin, 2007.

Pollan, Michael. *In Defense of Food.* New York: Penguin Books, 2009.

Vigilante, K., and M. Flynn. *Low-Fat Lies, High-Fat Frauds, and the Healthiest Diet in the World.* Washington, DC: LifeLine Press, 1999.

Acknowledgments

For funding the study on which this book is based, I want to thank the Susan G. Komen for the Cure Foundation. They encouraged unconventional protocols when the prevailing advice was that low-fat diets were best for health and weight loss.

I am grateful to the women who participated in the study and appreciate their trust in me. They committed to something that was contrary to what they thought, willingly changing their diets, keeping records, and answering my numerous questions. I especially want to thank those who told their stories, an unpleasant task that they agreed to do to help others faced with the same prognosis.

I asked Nancy Verde Barr to write this book with me based on our previous brief work together and her writing. Not completely knowing my nutrition philosophy, she took the material I gave her, researched information, and asked me questions that improved the material enormously. She made my dry, scientific writing readable—bringing clarity, logic, and humor to it. Nancy has made this a much more enjoyable book than anything I could have written.

Thanks so much to my father, Robert Flynn, my sister, Martha Flynn Tasca, her husband, Tom, and my nephews, Peter and James, for their help in trying countless recipes. They never complained when trying food "experiments," and offered constructive criticism that improved many of the recipes.

Thanks to our editor, Katie McHugh, for considering this project worthwhile, and to our agent, Jane Dystel, of Dystel & Goderich Literary Agency, for bringing it to Katie's attention.

—*Mary*

From the moment Mary told me about her work with diet and breast cancer, I was fascinated. When she told me that the success of her diet was due to the high use of olive oil, I was compelled to write about her findings. And so, I am first and foremost grateful to Mary for entrusting

255

her work, her theories, and the stories of her participants to my pen. If there are faults in the interpretations, they are entirely mine, not hers.

Testing recipes can be routine—or a heck of a lot of fun. I had a heck of a lot of fun because my cotesters, Tom Mainelli and Noreen Andreoli, made it so. I have worked with Tom off and on since 1990, when as manager of a local Williams Sonoma he hosted a book signing for me at which he served my recipes—and prepared them exactly as written (which, woefully, does not always happen). Between various stints in the food, wine, and olive oil businesses, Tom has always come to my assistance when I needed it, and for that I am beyond thankful. Besides providing precise attention to detail, Noreen generously offered her gloriously equipped kitchen for our work, and the space made the testing pure joy, especially because Rex, Noreen's affable Cavalier King Charles Spaniel, was constantly in attendance cheering us on. Well, mostly he was encouraging us to drop more food on the floor, but he was always cheerful about our presence in his domain.

I love writing, but when I am in the thick of it, I tend to resemble a deranged hermit—tethered to the computer days on end, same clothes, no makeup, hair uncombed, ignoring the phone and e-mails. A decidedly unhealthy way of life, which negated the benefits of all the olive oil, vegetables, and whole grains I was consuming. For saving me from the hermit me, my thanks go to Shay Lynch, who taught me the simple secret to his own full, rich, and varied life—balance. One of the ways he prodded equilibrium into my life was with frequent, exhausting, and stimulating squash lessons—the game, not the vegetable. Shay had no problem moving me around the court for more than two hours, after which I was energetic and eager to get back to writing—and balanced. He is, and always will be, dear to me.

It would be of little use for me to love writing if I didn't have an agent like Jane Dystel. She has always given my book ideas careful attention and valuable suggestions. She's the best and I thank her—as I also thank our editor, Katie McHugh, for believing in this book.

As always, I am grateful to Philip, Brad, and Andrew Barr for everything, but mostly just for being.

—Nancy

Index